# THOSE MAGNIFICENT SHIRES

# THOSE MAGNIFICENT SHIRES

S R Clark

ATHENA PRESS
LONDON

THOSE MAGNIFICENT SHIRES
Copyright © S R Clark 2006


All Rights Reserved

No part of this book may be reproduced in any form
by photocopying or by any electronic or mechanical means,
including information storage and retrieval systems,
without permission in writing from both the copyright
owner and the publisher of this book.


ISBN 1 84401 584 X



First Published 2006 by
ATHENA PRESS
Queen's House, 2 Holly Road
Twickenham TW1 4EG
United Kingdom




Printed for Athena Press

*In Memory of:*

*Sergeant George Edward Ellwood*
*512499, 114 SQDN, Royal Air Force*
*Who died aged 29 on Tuesday, 29th October 1940*
*Sergeant Ellwood, son of George Etherington Ellwood and*
*Ada Emily Ellwood, of Asselby, Yorkshire.*

*Remembered with Honour*
*Runnymede Memorial*

*Commemorated in Perpetuity by*
*The Commonwealth War Graves Commission*

*About the Author*

S R Clark is farmer born and bred.
Having said that, there is not much more to be said.
He has never been to London city;
Some folk would say 'What a pity!'
Not far from his home will he ever stray;
Still, he knows about things much further away.

## *Acknowledgements*

I would like to express my utmost thanks to the following people:

Many thanks to Mr E J Barton for his assistance in aviation research. Also the late Mr Morris Parker and the late Mr Harry Drury for the loan of photographs.

A special thanks to Mrs Audrey Ellwood of Warren Farm, Barmby on the Marsh, near Howden, and family for the loan of photographs, and especially for the Debt of Honour document in memory of Sergeant George Edward Ellwood, their very close relative, who is forever in their hearts.

Kind regards to Mrs Briscoe of the Doncaster and District Family History Society, and Jennifer Johnson of East Yorkshire Family History Society, for their help with research into family and local history.

Much appreciation for help from Mr Stephen Botcherby with typing the manuscript.

And to all the very fine people I have known, congregated and worked with, and all those whom I have taken the liberty of mentioning in my manuscript.

S R Clark

# *Preface*

These memoirs are a true account of my childhood and early working experiences with farming people. The people are real. Their triumphs, disasters and tragedies are all revealed, including those of my own family and our ancestors.

It spans a working life of over thirty years, from humble beginnings working on a small farm at the age of fifteen years, to being made redundant from a large group of farms, comprising thousands of acres, when in my fifties.

It is hard to believe that life in the small East Yorkshire village of Asselby, where I lived for a decade of my early childhood (1935–1945), was before television, videos, penicillin, polio shots, antibiotics, frisbees, fluorescent lights, credit cards, ballpoint pens, pop stars, computers, listeria, salmonella, and so on.

However, it is so, and I have many happy memories of those days in the quiet little village with a population of one hundred and seventy-seven people, mainly farmers, farm workers, smallholders, and others associated with the farming community of the day.

Everyone knew each other well, but at the same time were very independent and got on with their own business, whilst maintaining a friendliness and communal spirit, so often sadly lacking in our lives today.

No one worked on the Sabbath, except for livestock tending. Almost everyone attended a chapel or church, at least once, and in some cases twice, on the day.

We had plenty of entertainment, which we made ourselves, but alas our days were darkened in more ways than one with the outbreak of the Second World War in 1939. It was a very tense time for everyone, with friends and relatives taken from them, not knowing whether or not they would return. However, I am pleased to say that most of them did return safely.

However, it has to be said that the war put farming people on a better financial footing than they had ever known before. No one made or had much money before the war.

Farming had been in a bad financial state for years, and as a result farms were easy to obtain, to either rent or buy. It was very much a case of he who dares, but may not always win. After all, farming was the only industry in our particular village, and the boost to the industry came from governmental financial support, in order to help farmers and growers produce more home-grown food, and thus help greatly with the war effort. The 'Dig for Victory' campaign slogan became widely adopted by farmers and gardeners. Farmers were required to plough up old grasslands (some of it never having been arable land in living memory); others dug up lawns and sports fields. Most of this new ground cropped well, but was not without its problems, such as poor drainage, pests and diseases.

I have no doubt these particular war years were the turning point and the start of the golden years for farming, and have helped to create our fine agricultural holdings and estates, which Britain enjoys today.

Needless to say, village life changed after the war too, perhaps for the better; but the humble pre-war years never stray far from my memory, and the happy childhood – yes, even during the war years as well, with all its rationing and shortages of just about everything.

When one compares those days with our luxurious lives nowadays, we cannot envisage how we survived, not having any electricity, central heating, sanitation, and all the mod cons of today. It was by making the best of what we had. Almost every cottage and house had at least a small amount of land attached to it, which was used to grow vegetables and salad crops; back yard fowls kept us in fresh eggs, and also other livestock such as ducks and geese provided meat and eggs too. Most people killed and salt-cured pigs. 'We lived plain but well' – as a true Yorkshireman would put it!

Unfortunately, the trend over the last decade or so has caused us to become money-oriented, and surplus land around a property is now often put up for sale. It is very sad indeed how landowners have crammed all the new properties into what used to be lovely, well-spaced old villages. I suppose we have to put it down to progress, and I appreciate we have to move with the times; but at what price, I wonder.

## *The Shire Horses*

Oh, those magnificent, beautiful animals! Yes, I can remember every one of them. They are etched in my mind for ever. I often picture them in my dreams as I lie in my bed. They walked so proud with their large shod feet, clopping down the streets. They were very intelligent and knew their own yards and fields and virtually the time of day.

As the young ones were being broken in they quickly learned what was expected of them and very rarely gave much trouble. I cannot pretend to remember much about them when I came to live in the village of Asselby with my parents during 1935 at the age of two years, but from then onwards they were an inspiration to me.

I was, in fact, afraid of them from my earliest memory; but as time went by, became confident enough to approach them closely, and later I rode on their broad silky backs. There were approximately fifty in number in that village during the decade 1935–1945, plus foals and growing stock — mainly Shires and a few other breeds.

The farmers were, in fact, totally dependent on them to work their farms. Although tractors were very gradually beginning to appear by then, the best horsemen showed very little interest in them, and the farmers saw them as a threat to soil structure, and creating a multitude of other setbacks which proved to be short-lived.

Mr Richard Fletcher farmed East End Farm with five horses, namely Peggy, Farmer, Bonny, Punch and Tidy. Tidy was a large old brown mare, the mother of Peggy and Farmer by a blood stallion. They were half-legged as a result of this and worked as a pair; they were very keen and versatile.

Farmer, in fact, lived and worked all his life in Asselby. Mr Fred Harrison bought him when Mr Fletcher sold up in 1940; he worked on until the early Sixties. They looked very smart on the sale day, all spotlessly groomed and spruced up. Bonny was a dark brown mare with a very kind nature. She and Peggy were in foal to Mr Hatfield's Shire stallion for the sale. Bonny bred two foals by a Percheron stallion; they were broken in and sold as two- and three-year-olds but they never worked at Asselby. By being dapply grey they favoured their father: such a beautiful pair of horses. The horse stable on these farm sale days was the centre of attraction. The horsemen made a very great effort to make the animals look their best and show them well before the auctioneer. Punch made ninety guineas; I recall it being the topical piece of conversation in the village at the time.

The Everatt brothers farmed East End House Farm with six horses whose names were Captain, Shot, Blossom, Smart, Dick and Darling. Captain was a red Shire; a really big fellow, he showed some dapple in his coat in the springtime with his new hair. He looked very handsome but he could not wear a conventional collar, and always wore a breast collar. Perhaps he had rather a tender skin, but I cannot recall the reason. Shot had been his partner but he developed a very messy disease called grease, which causes the legs and feet to swell enormously. They constantly oozed very smelly pus, and the horse lost its condition, perhaps due to pain and discomfort. Mr Everatt tried very hard to cure him and he was isolated from the rest. I remember the village horsemen discussing this and saying they would be lucky if it did not spread through the other horses in the stable. This was not so, but poor Shot never really recovered, although he did work on for a year or so doing the odd jobs.

Blossom was a grey Shire brood mare; she became so broad when heavily in foal that she could not pass through the stable door, and had to be kept in a loose box. Smart, a black Shire, was also a brood mare. These two worked as a pair for years. Darling, another bay Shire but barren, worked with Dick, a chestnut gelding with a silver mane and tail. He was the only chestnut gelding in the village, a small, shy little horse. He plodded on year after year, 'never sick nor sorry', in the old horseman's phrase.

Mr Lapish farmed Croft Farm with two to three horses, namely Kitt, Star and Gypsy. Kitt was a small, half-legged brood mare, the mother of Gypsy, Prince and Duke, which Mr Lapish broke in and sold. This was his system. Each year he would break a gelding or filly to sell, as he did not need them all, being a small farmer. He was very highly skilled at this; he had endless patience and was fearless of these wild young animals. Star, a brindled barren mare, was very keen. He used to say she was awkward to work with another as she seemed to be pulling it along all the time, but in a cart or any other single horse job she was tireless. Gypsy, a very pretty little black mare, was by a black stallion that had four white feet and left some very nice horses in the area; he was owned by Mr Hatfield of Brind, near Howden. When Mr Lapish sold up and retired in 1944, Gypsy was bought by the Johnson family who farmed at Balkholme. Star was also bought by Mr F Johnson of Asselby (no relation), and worked on in the village for some time.

Mr J Barker farmed Ashgrove Farm, Manor Farm and Home Farm. Ashgrove Farm was later bought by Mr F Heseltine, and he farmed it with two horses, Jim and Cobby. Jim, a large brown horse, was very nervous. In fact, no one could work him but Mr Heseltine, as he broke him and used him from the start. He could be a very perilous animal to those who did not know him but could pull almost anything one yoked him to; he had the strength of an elephant. Cobby, a grey Shire and a younger horse, looked tired out after a day's work alongside Jim, but was raring to go again the next morning.

*The Shire Horses*

Mr W Everatt — no relation to the former — farmed Box Tree Farm, with in all six horses, namely Boxer, Captain, Prince, Bonny, Duke and Brisk. Boxer, a black horse, was in fact then in his old age, and soon disappeared, so I cannot say much about him. Captain, a very large red Shire, again with this lovely dapple in his coat in the spring, worked with Duke, a bay Shire. He was one of a matching pair, bred in the village by Mr N Shaw of Eel Hall Farm, who sold up in 1931. Prince was also bred in the village by Mr V Clayton. A very fit and keen gelding, he did in fact run away in a cart one Good Friday. This was bad news to a horseman, as a rule, for once done, it was feared the particular horse would do it again, and become unreliable and need constant surveillance. Fortunately, he settled down and worked for many years as a good reliable horse. Bonny, a pure-bred Shire bay with black legs, always looked in perfect condition. Brisk, another dark roan mare with a blaze down her face, walked and stood so proud. These were a team of really excellent horses, but Mr Everatt never went in for breeding.

Mr J Barker farmed his three farms with five or six horses, namely Tom, Tidy, Jet, Bonny, Boxer and some others. They always seemed to be buying and selling, therefore I cannot pretend to know much of them all. Tom, Tidy and Jet were old, worn-out horses. No doubt they had been competent in their best days, but in spite of being slow they plodded on in the harvest carts and similar easy jobs. Bonny and Boxer were two young ones they bought just before Mr Barker retired. Boxer was a dark brown Clydesdale horse with four white legs and Bonny was a Clydesdale mare, strawberry roan; she had a lovely coat. They were a very smart pair.

Mr Palmer farmed Crossings Farm (which was only a small farm) with two horses named Topsy and Ginger. Topsy, an old red Shire, and Ginger, a pure-bred Suffolk mare, worked this small farm year after year giving no trouble at all. Mr C Bolden was able to manage most of his farm work with one

## The Shire Horses

horse. His farm at the west end of the village was a small one, West End Farm. At the time of my memory he had just the one mare, called Tidy, a Clydesdale.

Mr F Johnson farmed Field House Farm with four to five horses, whose names as I recall were Dick, Darling, Star, Boxer and later, Mr Lapish's Star. Dick, a dark tan Shire, was a plodder, one hundred per cent reliable. He worked on day after day, year after year, and never seemed to age or change at all. Boxer, a very large black Shire with four white feet, was an outstanding horse. It was said he was the best-looking one in the village, but for a slight defect. He had a stiff front leg which caused him to walk and stand not quite his maximum height, but the speed and strength of this horse was very outstanding. However, he was a little nervous, and did not like working on his own, which could not always be avoided. Therefore, he had to be handled with respect on these occasions. Star, a similar coloured brood mare, worked with Boxer as a pair. She was keen and a good all-rounder, perhaps handicapped being a brood mare. These mares, after a little rest from foaling, were put back to work. This caused mother and foal some distress, having to be parted for several hours per day. The mares, in particular, were uneasy and had to be handled carefully. I can recall my brother and me feeling very sorry for the unhappy youngsters crying for their mothers; even our presence made them feel a little better, but we were accused of spoiling them and sent packing. Darling was an old bay Shire and we nicknamed her Granny, but looking back, I can see this was not very apt, as she was barren; but she was a very reliable old plodder nonetheless.

Mr G Ellwood farmed Eel Hall Farm with five horses, namely Prince, Captain, Violet, Smart, Bonny and young Billy, who sadly fell for a double tragedy. His mother, Violet, an iron-grey Shire, died from some complications after he was born. However, one of Mr Ellwood's sons, Percy, brought him up on sweetened cow's milk, and he made a fine young roan colt. Alas, shortly after he was broken in for work, he was struck by lightning, as a result of standing under a pink hawthorn tree. At least this is what they suspected, as the tree was clearly damaged and in fact also died. Young Billy was not killed outright, but severely injured, and could not stand on his legs anymore. After an examination by the vet, it was confirmed he was permanently injured. This left them with no alternative. He was sadly put out of pain with a bell gun in the paddock he had happily grown up in, in spite of being an orphan. He was a pleasant, good-natured little animal, and it broke many hearts to see him hauled into the knacker's wagon – such a tragic end.

Captain was a yellow bay Shire bred by Mr Norman Shaw, whom Mr Ellwood followed on Eel Hall Farm. As aforementioned, Mr Everatt bought his double, Duke. These were a matching pair, and I would say the best pair in the village but I cannot pretend I ever saw them work as such. Prince, a purebred Shire, was more like an elephant. This is what Mr Ellwood's sons used to say. I am afraid it has to be said, he was positively ugly, but as to work, one could not tire this old tyke. I recall Percy, the elder son, saying he banked up a six-acre field of potatoes with him in an afternoon without stopping once, and this was damn good going. Smart, a chestnut mare, worked along with Bonny, a grey mare. Both were barren and plodders.

Mr Fred Harrison farmed Back Lane Farm with two pairs of horses, namely Prince, Bonny, Farmer and Metal. Prince was an old Shire, dark brown in colour. Fred used to say he always looked as though he had 'eaten his bedding'; in other words, whatever one gave him to eat, or how much of it, he never altered from his frail looks, but he was also the tireless type. Bonny, a black Shire, died after a hard day's work, quite unexpectedly. Farmer was a cast-iron horse, another one bred and bought in the village. This was always good practice, as one knew the horse's form and ability. Old Farmer had it all, and lived to be over thirty years old. Metal, a brown mare with black legs and feet, was very keen; Fred bought her to replace Bonny. Old Farmer seemed to know he had got a new partner similar to himself, and the sight of these two speeding up and down the fields pulling a plough was a remarkable sight.

Mr V Clayton farmed Sycamore Farm with his six horses whose names were Gypsy, Boxer, Jerry, Tidy, Judy and Bonny. Gypsy, a light brown brood mare, bred a lot of good foals by Mr Hatfield's stallions. Most farmers kept at least one brood mare, but very few kept their own stallion, because these animals could be very difficult to handle and, in fact, dangerous, if not shown the utmost respect. Consequently, these stallion keepers took the job in hand by touring the farms and villages during the breeding season, often by appointment. This was referred to as 'travelling the stallion'. One would often see these stallion travellers parading the cattle markets, canvassing for trade.

Mr Hatfield, senior, would travel a black stallion, and Mr Hatfield, junior, would travel an iron-grey stallion. These colours were usually the most sought-after. Boxer, a dark Suffolk punch was a real beauty, perfect in every way. I have to confess I learned two things about him I never knew until last year. During a conversation with Mr Dick Clayton, a very competent horseman, he told me Boxer came from Leeds. He had, in fact, been a coal-dray horse in his young days. While I could remember him getting foot canker — for which, unfortunately, there was no cure — and the Clayton family being upset about it, I did not know he was put down at home before being taken away. Dick said he felt he knew where he was going that day, and Boxer could not be subjected to any ill treatment at all.

He could not bring himself to tell us young lads at the time. Jerry, a dark brown gelding which they bred, was in fact named after my younger brother, Gerald, as the Claytons always called him 'Jerry'. Tidy, a blue roan Clydesdale mare, was like a painting, so perfect in every way. Judy, a strawberry roan Clydesdale, was very similar to Tidy with a silver mane and tail. These two young mares made a perfect pair, and had they been the same colour, they would have been of show quality. The Clayton brothers made great efforts to keep their horses in great shape and it showed as they proudly trotted them through the streets of Asselby. Bonny was another nice filly that they sold for some reason, only to find out later she had gone to a bad home, which they were none too pleased about.

Mr H Stead farmed Riverside Lane Farm, again quite a small farm with only two horses, whose names were Captain and Bonny. Captain was bought as a yearling and later broken in to make a fine big red gelding. Bonny, a black and white mare, was a very good all-rounder.

Mr E Hutton, a market gardener, kept an old mare called Diamond for light work, and later he bought a mare in foal at Mr Holland's sale at Barmby on the Marsh. She had a fine iron-grey colt, which they broke in to replace the old mares. He was quite a handful too, as they did not have enough work to keep him calm at times.

I feel these animals should have a place in history; they were so loyal. They did their bit towards keeping that particular patch of Britain's heritage tilled and tidy.

They hauled large loads of farm produce in all weathers. They came when called from their lovely green pastures for yet another day's hard work and toil. The harvest time was a very hard spell for them, particularly hauling the binders cutting the corn. I have seen the poor beasts with balls of sweat running from them and the flies pestering them; it was cruel. These machines took at least four horses to operate, with plenty of rests and a change if possible, but I am afraid they were sometimes driven on mercilessly in a bid to beat the clock and the weather.

As the tractors gradually took this job over from the horses, there came a turning point in mechanisation. They speeded up the job greatly. While the early tractors were very basic and merely hauling machines, they became more versatile, i.e. with hydraulics and PTO drive. Now this was really something, as the original binders' mechanism was driven from a land wheel, which when operated in soft conditions or laid crops caused the whole machine's momentum to slow down, blocking the whole thing up solid sometimes with damp material, which had to be tugged out by hand or hand-cranked through before one could proceed on again. The power-driven machines cut out these problems.

During the late Forties and early Fifties, we had that period with both horses and tractors, which my late father always said should have been kept up. I tended to be in agreement with him, for there were jobs which one could do cheaper and more efficiently than the other, and vice versa. Unfortunately, the trend from the horse rapidly took over, and they quickly became redundant.

Finally, I am sure we owe a lot to the people who struggled on to keep the heavy horse breeds alive, as they became very low in number in the Sixties. The Shires also lost some of their features and characteristics. This was the main issue in the show rings amongst the few remaining enthusiasts, while the Suffolk and the Percheron hold their own types, and can therefore be regarded as unique.

I suspect there have been times when the Shire and the Clydesdale breeds of horse have been rivals to each other due to the breeders' beliefs, their stubbornness and their desire to dominate and perpetuate bloodlines... not unlike the tempestuous merger of Harry Ferguson and Henry Ford with their tractors, resulting in the famous 1948 lawsuit between them. This, however, was a much more expensive contest than the animosity between the horse breeders, which remained on a more civil basis.

Take, for instance, the strategy of the dedicated and successful Scottish Clydesdale breeder, Lawrence Drew. He blatantly refused to cooperate with the newly formed Clydesdale Horse Society in 1877, and similarly influenced many other breeders to do so. He firmly believed that the two aforementioned breeds, after several years of interbreeding, had resulted in there being no genetic or characteristic distinction between them. A strange philosophy, I would have thought, for despite this, Drew during 1883 founded a select Clydesdale Horse Society. Again, such a move seems to me to contradict his own beliefs, or perhaps he had compromised – but on his own terms. However, with all due respect to the man, it looked as though he was hell-bent on domineering the Clydesdale camp, had he lived long enough.

Another fact that statistics reveal and perhaps might also confirm his former beliefs is that a Shire stallion of outstanding repute was sent to Scotland in 1866 to serve scores of Clydesdale mares, which shows no one was over-fussy about bloodlines at that time. When I visualise the horses that walked the streets of Asselby during the 1930s, I see the features of many a 'Harold, 1881 to 1901, Shire Sire Supreme', 'the great-grandson of Harold, 1903 to 1924', ditto, 'Vulcan, the Earl of Ellesmere's Shire stallion, 1883 to 1903' – another handsome fellow who was given the chance to sow his golden seeds in Scotland.

The Earl of Ellesmere was the instigator of the forming of the Shire Horse Society, resulting in these fine true Shires. However, I fear should he be here today, he would be disappointed about the short thickset features which I see in the portraits of the former Shire stallions. These indicate the absolute pure Shire, unfortunately no longer to be found, which highlights Mr Drew's relevant theory; and likewise he would have no problem convincing me of the validity of his 1877 theory regarding these animals.

In 1940, at a demonstration featuring the Ferguson system, Harry Ferguson spoke from his rostrum, and in front of many senior government officials, and umpteen other interested bodies, saying that 'a million man-hours per day were being wasted tending horses, and an equal amount of man's time is wasted growing food for them'. Such unbidden words revealed a dedicated businessman with much more empathy towards the potential millions of pounds he envisaged making from the launching of the Ferguson system, than on the post-war food shortage, or the economics of the farmer. And his words, and indeed his invention, neither said nor did anything in favour or recognition of the ever faithful and wholly reliable workhorse, which had been the major source of power to British agriculture for at least two hundred years. No doubt he meant well, and there was no way he could have envisaged that his labour-saving mechanisation crusade would lead to today's utterly absurd and tragic man-made scandal, which our agricultural system has lapsed into.

I refer to the incompetence of the boffins presiding in the Euro Palatial Debating Chamber in Brussels, costing the European taxpayer £6 billion per year. The British politicians are equally as crass, plus the bloated common agricultural policy (CAP), which rewards farmers for growing nothing, adding hundreds of pounds to the average family's food bill yearly. I maintain that the post-war food shortage was never that serious, nor was it down to the slower pace of the horse-farming era.

Food shortage, food mountains, slump, glut, supply and demand have always formed an integral part of agriculture. One has to ask the question, wouldn't the whole issue of agriculture function better without being subjected to constant meddling by the politicians?

## The Horsemen

It would be very unfair to name the best horseman in the village, but as in any other team there is always a star, and he existed here too. Having said this, they were all very skilled people, and these smaller farms were where a lot of them learned the basic rudiments of horsemanship. They would start as day lads, perhaps as early as fourteen years of age. After being taught how to gear up a horse for any particular job, the lad would almost certainly be set off with the quietest old horse yoked in a cart doing the simple carting jobs. His progress would be observed by the foreman and the head horseman, and then when the time was right he would be set off with a pair doing the more skilled work. Each man would normally work the same pair of horses most of the time. The horses were worked in selected pairs, because one achieved a better performance from two horses that were used to each other and the same handler. Of course, each county had slightly different ways of driving or handling their horses.

'Driving' was the general phrase used for a person yoking two or more horses to an implement, cart, wagon or machine and making them go where one required. Again, this was a skilled task which required a lot of training and patience. When operating a pair of horses in this flat part of Yorkshire, one would be known as the 'string' or 'line' horse, and the other, the 'far side' horse. The line horse would be the leader, or the one that took the driver's orders and line signals. The far side one would be tied with a line from its bridle or bit, namely a false line, just far enough back from the other's inner trace to restrain it, and thus keep both horses pulling together. A three-horse team would be driven in the same way.

Technically, one was driving two or three horses with one rein or line. The horses were also given a lot of verbal instructions, and they soon learned the meaning of these various words. A good handler would keep talking to them in a quiet and regular tone. The horses quickly detected an erratic mood, and they would become so too; but at the same time one had to be firm and positive, for they could also detect whether a driver was in full control, or not. In the latter case, they would take advantage and enjoy it.

A good line horse would not go very well at the far side; it would become confused and in some cases irritable and impossible to handle. This brings such a horse to my mind – Mr Frank Heseltine's horse, called Jim. Now, he was strictly a line horse, and Frank had broken him in and used him from the start. He was very nervous and headstrong, and no way would he go at the far side. He had to be the leader, but he was very good at it, and could walk down the fields as straight as an arrow with his head held high, looking so proud and confident. The performance of that horse was really outstanding, and despite his extra large frame and fierce appearance, he was good-natured and safe; but no one could make much of him except for Frank. Now, I refer to a very competent horseman here; he was a man with cast iron nerves. There was nothing he liked more than to be in combat with one of these wild young horses. He could soon get it through to them that he was in charge, which was of paramount importance during a young horse's training.

Basically, all horses were broken in by the same procedure; therefore it could be assumed that they were a uniform lot. Not so. Loosely speaking, they fell into three categories: a line horse, a far side horse, and one that would go anywhere. The latter would almost certainly be an aged, quiet horse with a lot of intelligence and experience. These attributes would be discussed between the vendor and the purchaser during the negotiations of a deal regarding a particular horse, and would be honoured as near as was possible, for obvious reasons.

Of course, the horseman's task entailed more than driving and managing his horses. He had, when working the land, the additional task of setting, adjusting, steering, watching and maintaining the implement or machine they were hauling. Ploughing and ridging were the major tasks. The ploughman had to be accurate when marking out his fields into thirty or forty yard lands, and have the ability to turn a long straight furrow with only a fraction of an inch deviation, and so keep each bout of ploughed and unploughed land parallel. As the horse ploughmen walked behind their ploughs, one could see them constantly watching and studying them. I know, from the small amount of horse-ploughing experience I have, that it gives one an enormous amount of satisfaction and there is such a lot to be learned from turning your first furrow, to becoming a competent ploughman; and this was a must, if one was ambitious and keen for promotion. Many horsemen were quite contented with the job, and were happy to spend all their working life in it.

As mentioned earlier, the ones who were interested in the farming ladder usually moved on from the smaller farms, which we had in Asselby, to ones with more opportunities. For example, Sand Hall farms, as owned by Captain E P Schofield's establishment; Boothferry and Court House farms, farmed by Mr R L Walker; Percy Lodge, farmed by the Tinsleys; The Groves, farmed by Mr Patchet; Cavil Hall, the home of Mr Stobard; Burland Hall, farmed by Mr C Brown; and Decoy Farm, the home of Mr H Firth, to mention but a few. These were some of the farms horsemen aimed to get a position with, and they would feel highly honoured having achieved such employment. These farms and many others were widely talked about amongst the farming people in the villages around Howden. Particularly in Howden, which was a very popular and busy little market town during the farming days, the farmers would congregate in the pubs in the evenings and on market days, exchanging their tales and beliefs and progress on their farms, etc.

There were a lot of travelling people involved in farming too. They brought and carried news as they moved from farm to farm and village to village going about their business.

There were the threshing contractors, the stallion travellers, the pig killers, and an old man named Mr Henderson of Goole. He cycled far and wide selling horse medicine and conditioning powders, and all things relating to the well-being of horses. Another character, also of Goole, by the name of Tinker Abbey, took on, by price, the tarring of Dutch barns and other farm buildings. He was a white man but one could easily mistake him for the exact opposite after a day's work. He would often spend an hour or two in the pub at midday, then back on the roofs again. He never used a roof ladder and must have had nerves of steel to operate the way he did. The farmyards were the centre of attraction in those days; there was always something going on.

Mr Tom and Ben Drury of Howden were also two regular visitors. Looking back, I often think they should have been called Morecambe and Wise, because there was constant laughter when they were around. They were known as hay cutters, not as it grew, but from the large stacks of hay and clover they cut, and pressed this loose fodder into low-density bales, or trusses, as they called them. This was very hard work. All by hand, they cut these stacks into squares, with a razor-sharp hay spade or knife, carried the material down the ladders with a large needle through the middle of the section to hold it together, and placed it in the press, a piece of apparatus similar to a guillotine used for the job. Tension was applied by hand, pulling down the press handle, with the material between two wire threading boards. When the wires were threaded and tied, the tension chain was released, thus leaving a nice, tightly packed, sweet-smelling package of fodder, usually for sale at a destination, which I will come to later. Again, they had to have some very sharp knives to do this job. They also had some safe operating rules which they constantly reminded each other of, but despite these rules Fred did have a very nasty accident, which according to Tom (who was his father, by the way), was his own fault. He had carelessly put the trimming knife down. He later trod on the handle, causing it to spring up and gash his leg. Fortunately, this happened in the mid-Fifties, when Tom was a retired police sergeant, and had some first aid knowledge. He therefore acted quickly by doing what he could for Fred there and then, and later drove him to Goole Hospital. Had it been in the Thirties, when us lads used to watch them at work, and Fred and his previous mate travelled on old cycles, things could have been much worse. Needless to say, Fred got no sympathy from his father for allowing it to happen in the first place.

To get back to the horsemen: again, Howden was where most of them would negotiate over the work vacancies. There were no formal interviews in plush farm offices with frustrating 'let you know' results. They worked largely from the old proverb, 'A Yorkshireman's word is as good as his bond.' The farmers, and in some cases the foremen, would meet in Howden marketplace, and the pubs around it, during Michaelmas week, the third week in November. This was the time when the horsemen renewed or changed their positions. They were employed or hired, as they termed it, on a yearly contract. Once an agreement was reached between them they shook hands on it, and in 99 per cent of cases it was honoured by both parties.

These contracts included board and lodgings, but the wages were poor, and varied according to status. A top horseman during the early Thirties would receive around thirty pounds per year. There were no set hours, but they were generally well catered for and made to feel at home. The larger farms would keep six or seven pairs of working horses — even more, counting all and sundry. They were operated with a very strict regime and the head horsemen were treated with respect by those below him in status. He very often had his own stable, which was out of bounds for the other farm staff. Without a good reason for being anywhere near it, they would be in hot water. While the foreman's word would usually supersede the head horseman's, they would liaise together as the two senior men, and set a good standard of moral conduct, pride in the work and, above all, punctuality with their men.

The foreman, if employed with a gentleman farmer, would be ultimately responsible for the smooth running of his farm. To arrive at this position he would have to be the senior man with several years' experience as head horseman, and hold a good practical knowledge in horse breeding, feeding and keeping, etc. He would be able to stack, thatch, and be well versed in all the labouring skills. Should there be any technical problems with an employee, and a particular job that needed doing by a newcomer for the first time, he would then be capable of putting them to rights, and thus prove himself worthy of his job to them.

The men would expect this and respect him for it. Some of them were known as 'walking foremen', and this is literally just what they did. They would tour the farm on foot, organising and issuing orders in all directions. There would also be gangs of women workers, and Irish labour, to control during the busy seasons. The walking foreman would live in the main farmhouse or the foreman's house, provided this was part of his contract, and he would board the hired horsemen, other farm staff plus the household staff, the foreman's wife, and her staff. In the cases where the working farmer did not perhaps require a foreman, it was then usually his wife. In both cases, these ladies were very hard working and dedicated people, in spite of their work being so repetitive. All they seemed to do was prepare food, clean and wash seven days per week. They too had to work strictly to the clock, to coordinate with the men's timetable.

*The Horsemen*

As the horsemen were up and in the stable by 5 a.m., their horses were given their first feed, mucked out and groomed, and then it was back to the house for breakfast. The horses were the obvious reason for the men needing to live on the job. They required a lot of attention before and after their working day. The horses were geared up and off to the fields by 6 a.m., and the horseman off the ladies' hands for a while. There is no doubt in my mind that having horses around the farm makes for better human relationships. The horsemen lived for their horses. They were very proud of them, as they brought them out of the stable, all immaculately groomed and geared up with martingales and other brasses and bells on their collars and so on. Yes, there was a lot of work in the stable, especially in the winter months when the horses were sleeping in. They would be covered in mud after a day's work if the weather was wet, and the lanes and field entrances would get in a very bad state; but at times they had to keep on using them. The spoked and iron-rimmed wagons and carts really cut the lanes up in wet weather. I have seen them carting turnips, manure and potatoes to and from the fields for most of the winter. One can imagine the state the horses were in after working in such conditions, and how long it took to clean them spotless again, but this had to be done. Then they would be inspected by the farmer or the foreman before the men could retire to bed.

One of my long-time friends, an ex-horseman, Mr Harry Beardshaw, who is now in his eighties, and tackles a long daily walk around the lanes with ease, has so often told me about his days hired in farm service at Decoy Farm, Rawcliffe Bridge, which was then farmed by Mr Henry Firth in 1926. He recalls a particular night when he and his fellow horsemen were in bed around 10 p.m. Mr Firth returned from an evening out, and made his nightly inspection of the horses. He was shortly to be heard bellowing at the foot of the stairs that they had only half cleaned them, and they had to get up and clean them again.

*The Horsemen*

Mr Firth was a very hard, strong man, built like Mike Tyson, and a very good farmer. I had the pleasure of knowing him later in my life. He did, in fact, follow my grandfather onto Decoy Farm. Harry also recalls how Mr Firth used to shoe his own horses and inspect their feet. This was always carried out on a Sunday morning, which was supposed to be their day off; but really these men loved all this hassle. They had a lot of competition too from their neighbouring farmers' men, and would visit each other's stables in turn to compare and perhaps criticise their work and so on.

The larger farms would have a stable man to assist the horseman, and being an ex-horseman he would know what his duties were. He would prepare food, carry in the long fodder, keep the stable clean and tidy, help to strip the horses' gearing, and inspect it for any flaws etc. He would also keep a fire going in the tack room, since there would sometimes be horse rugs to dry as a result of a wet day at work. The horses that were on the road had to have frost nails fitted and removed each day during frosty weather. These were sharp and kept the horse on its feet when walking on slippery surfaces, but they had to be removed every night for the horses' safety. Then there were all those paraffin oil lamps to fill and the glasses to clean. I could go on, but generally speaking they did the tasks between them, and were a cheerful bunch of chaps with a word for everyone. Even today there are a few old horse stables still intact. If ever I come across one, I cannot resist entering. I stand and stare, as they are now so silent, with the cobwebs hanging, and perhaps the rain dripping through a neglected roof. It is sad, and perhaps for the better that the old horsemen never knew what happened to their stables, for they too would have been very sad indeed.

## The Farms and the Farmers

Asselby during those days was the middle sector of the Knedlington Estate which held most of the farms and properties comprising Asselby, Barmby on the Marsh, and Knedlington villages. It was a staple-shaped parcel of land, lying mostly between the Rivers Ouse and the Derwent, and bordering the Leaconfield Estate at Newsholme. I suppose the fact that it was a no-through road from Knedlington and therefore a dead end at Barmby on the Marsh helped to make it the quiet and peaceful little estate it was; and I suspect that even today, its tranquillity will not be spoilt by the ever-increasing volume of traffic that many rural villages have to accept and tolerate these days.

The estate, during the decade of years I refer to, was owned by a Mr Mortimer, and Knedlington Manor was the estate's headquarters and main house. I used to feel disappointed not having seen the estate in its full splendour, as it was often described to me, with its gamekeepers, gardeners, woodmen, butlers, coachmen etc., but unfortunately it was around the time when I was born when these fine estates began to decline and crumble, due to various factors. I would say the state of Wall Street, money and its availability and the wars, together with the Depression of the 1920s and '30s, were the main factors. The wars were costly. We all had to contribute to them, some in person, others in kind. I maintain money dominates people and things, and still believe it was a great loss when these estates broke up. They were good for people, the environment and rural communities. They offered plenty of opportunities to the beginner, with all their farms to let, which were easy to obtain without a lot of palaver.

*The Farms and the Farmers*

The estates that survived took away these opportunities by subsequently taking the farms in hand, and they have been farmed in large units ever since. I sometimes wonder how much further on we are from those times, after several decades of the agricultural boffins, politicians and scientists telling us how and how not to do it all. Agriculture seems to be back in the same rut, which does not say much for this bureaucratic meddling. Recession seems to be largely at issue here; the previous version of similar circumstances was called Depression, but generally speaking I would say these words pertain to a slump.

Having said this, I must add that I have noticed a totally different reaction from the victims of these difficult times, as I clearly recall the earlier farmers who came to grief. Some I knew by name, and others I met over the years in my career. They seemed to pick themselves up, and one would see them quickly re-employed as a horseman, stockman, or even a labourer and they seemed content, perhaps even relieved in some cases. These days, they might reach for the depression pills, the psychiatrist's couch, and even contemplate suicide. I feel the former were better survivors, and better constituted to face their troubles. They were dedicated to farming as a way of life. If they only 'broke even' some years, so be it. Provided they could find the rent, they happily anticipated the coming season, proving themselves loyal to farming, the oldest of the country's industries. Their main goal was not necessarily a large bank balance, and I dare say they had learned the value of the maxim, 'Put not your trust in politicians.'

These Asselby farmers, and indeed many others, were a provincial bunch of people, only interested in what went on in their village or county, paying very little attention to the goings-on in London, let alone Brussels. But they were proud; they were farming a section of the East Riding of Yorkshire, which was considered to be almost the premier agricultural county in England, and the largest. The farms were neat and well cultivated. The farm buildings and stockyards, hedgerows, dykes etc., were kept neat and tidy. They farmed from traditional knowledge and beliefs, passed on from their fathers and grandfathers, and there were no quotas, subsidies, grants, National Agricultural Advisory Service, Common Markets or the like. Free markets always work, and we shall see them return. Perhaps we had a slight post-war food shortage, but the way the farm leaders have tackled it has shown their absolute incompetence; being bogged down with a subsidy-sodden world presents a far worse dilemma. Quality seems to have gone out of the window, overtaken by quantity — but where is the logic in producing mountains of sub-standard food? Then dumping it on each other, thus creating bad feeling and chaos amongst the whole community.

It is not a fallacy to say these pre-war farmers produced very excellent quality foods. Take the beef and the pig meats: one cannot compare today's artificially produced replicas with those lovely large hams and bacon sides, produced from 30- or 40-stone pigs, which were salt-cured and dried prior to using. Now that was the real McCoy! Older, larger, well-matured animals were the key to all this, plus a diet free from growth hormones, recycled animal offal, drugs, angel dust, or any other similar garbage, which produces an inferior product. Such modern produced meat is the cause for all the consumer complaints, now widely an issue. We now have people shopping around in search of a beef joint with taste, which we did not have when the beef was produced from turnips, mangolds, sugar beet tops, sugar beet pulp, marrow stem kale, hay, clover, linseed and cotton cake, home-grown cereals and straw, and so on.

I maintain this modern cheapjack junk feeding is utterly unnecessary — especially now that we have this set-aside farce, which helps nothing, perhaps only makes it more easy for the developers to get in.

This surplus land could be put to better use by growing

natural animal feeds, and setting up free-range stock and poultry units in place of the very unpopular methods of today. I must add, if I were a butcher buying a beast for slaughter in a market, I should be taking a close look at the stuff that falls from beneath its tail as it walks the auctioneer's ring, and beware of a yellow texture, for I know it was never that colour from animals fed on the former diet. Hence this could be a good guide as to what one ought to be buying.

No doubt these views of mine, on this particular issue, will be looked upon as very one-sided by many people, but we cannot deny quality of food is at this time a very controversial issue, which cannot be rectified with lack of collaboration from the manufacturing sector. Again I feel strongly about this — that it will never be rectified while our animal feeds contain scientific junk. Furthermore — believe it or not — I am not an old stick-in-the-mud. I do believe in progression. I am aware that science and technology have done some very beneficial work for agriculture, which deserves some recognition, but I do feel it is out of order in some aspects. Take the BSE scare, for example. Now, there is a man-made epidemic, with far more to it than the public are allowed to know, and I suspect a lot of its facts were swept under the carpet, hopefully never to be noticed or heard of again. But that, of course, remains to be seen.

I must also take a swipe at the genetic engineering boffins, who are posing a serious threat to us all. There is no justification for their latest venture of meddling with our fine breeds of farm animals. It is total and utter nonsense, which our government, the cattle societies, the unions and all other bodies connected, must take all possible action to block immediately. We have a small consortium of people with not a care for the potential catastrophic results they will be creating, but only in increasing their own financial gain.

However, to get back to Asselby. Mr Richard Fletcher farmed East End Farm, but he was not an Asselby man. This is how the Asselby octogenarians viewed a new face; nothing personal, just one of their whims, perhaps. Mr Fletcher followed the Kemp family on East End Farm in the late Twenties after they moved to Boothferry café. He came from Owston Ferry, and a long line of farming ancestors. He was a magnificent dark-haired gentleman, with manners and personality to match, and was also a local chapel preacher and Sunday school teacher. Therefore, us boys and girls knew him very well. He would become very uptight during Sunday school sessions if our conduct declined a little, but it was soon forgotten when he required us for some root crop weeding etc. I recall when he first grew sugar beet in 1939, he made the crucial mistake of sowing it after potatoes. It was a wet summer, and the crop became lost with volunteer potato plants. A gang of us singled and pulled them out, for four pence a row, and one had to work hard to do two rows per night; but we had the added bonus of our pockets being full of small new potatoes, as it was well into June when we finished them.

Of course, sugar beet growing was a total different kettle of fish in those days. It was drilled on ridges with a two row bobbin drill, at a seeding rate of between 6–10 lbs per acre, resulting in a thick row of seedlings which were then set out, or 'struck and singled out' as the operation was termed, all by hoes and hand singling to a crop stand. Sugar beet, being a relatively new crop addition, was accepted with some caution by these farmers, but once the technique of growing it was learned it made a good break crop and mixed in well with their present rotation. It also did well on that good-bodied brown sand land, and the warp land also.

Mr Jack Jarvis was the Selby beet factory 'fieldsman'. He toured the area in a Vauxhall motor car at about 20 mph, and spent half his time on the wrong side of the road, while staring at the sugar beet fields. Being a ladies' man, he spent more time chatting up the farmers' wives, than discussing the progress of the beet, and he knew all the current gossip between Selby and Asselby; but I dare say it was all a part of his job, which he did to the best of his ability, and he was a well-liked, cheerful chap with a word for everyone.

Mr Fletcher farmed East End Farm with two horsemen, Stanley Graham and Leslie Dennis, a lad called John, and an old labourer called Mr Hutton, who lived in Rose Cottage. The others lived in the farmhouse with Mr Fletcher and his housekeeper, Miss Nellie Oakes. He was also the first and only farmer to attempt making silage. They made it in a round mesh silo from a long mixed cereal crop, cut green with a reaper and carted and handled with forks. It was very hard work, but it turned out quite nice stuff for their first attempt. The other farmers received it coyly, as a possible rival to their 'roots', and it never caught on. Unfortunately Miss Oakes became seriously ill during 1942, and died at the age of thirty-five. The same year, Mr Fletcher was clearly beside himself as a result. She had been a very faithful and loyal employee to him. I can also say they were close friends. He seemed a very lonely man afterwards. He did, in fact, retire from farming, and sold up in the spring of 1943 and later moved away from the village.

In Asselby they were all mixed farmers with similar cropping patterns, perhaps differing slightly in methods and beliefs. The farms ranged between 50 and 150 acres, comprising a strip of warp land behind the River Ouse bank, running into a clay hollow, and then rising to brown sand around the villages. The marsh lands at the Selby side of the villages had, no doubt, been pastures for many years, as they were full of herbs, and made some lovely hay and were excellent for grazing stock. They never 'burnt off' during the hot summers. The arable land had obviously been expertly mapped out into fields, each one with its particular soil type, and with hedges for boundaries and dykes for drainage etc., making the fields easy to work, and offering a good choice for crop rotation, which was strictly adhered to. This was a practice learned from their forefathers, who had in the past been tied to covenants and expected to farm in with good husbandry, in order to ensure that the fertility of the soil should be kept up.

Continuous corn growing was virtually unheard of. Clover, corn, roots, potatoes and fallow were mainly varied. Clover was a valuable, widely grown crop, which I gather is making a comeback, mainly because of its ability to put free atmospheric nitrogen back into the soil eventually. It was a fodder crop, particularly for the horses, and there was always a good demand for it from the town horse-keepers, and the town intensive dairymen. Mr Spink of Hail Mill, Howden Dyke, was the local fodder merchant who, as previously mentioned, employed Fred and Tom to cut and press hay and clover for him. It was perhaps thought to be bad husbandry to sell clover off the farm for some, but they never sold any clover until their own requirements were catered for.

The warp land grew some excellent clover after two harvested crops, and there was a good one left to plough in, in the winter. It was undersown in the cereals for 'one-yearly', and sometimes created a problem during a wet summer which would encourage its tall growth. At harvest time, and once it was tied in the sheaves by the binder, it took much longer in the field to dry, and become ready to stack; but it did produce some palatable straw, amongst which the cattle loved to forage.

The Everatts were an old established family in Asselby. Mr Richard Everatt died in 1903 aged seventy-eight, and his son, John, died in 1937 aged eighty-three, leaving five sons and some daughters, whom I cannot pretend to know much about. But I do just recall the old man walking very slowly to the fields each day with a hoe. He would spend hours chopping thistles in the corn and other crops. He did, in fact, fail to make it back home one day, and was found collapsed by the roadside − not dead, but it was the beginning of the end for him. He had been a very good farmer, absolutely dedicated to farming. He was known in his young days as the potato king, as he was highly skilled in growing potatoes of the best quality. Mr Tom Walker, who had been his head horseman many years ago, told me of his very outstanding knowledge and practical ability to farm to the highest standard. Therefore, it was a sad loss to East End House, their home and farm, the family and lots of other people when the eminent old gentleman passed away.

Fortunately, he had left a good crop of well-trained, equally dedicated sons to carry on the family tradition. Richard, the elder son, had already left home and started out on his own on a farm at Knedlington, and later moved to Barmby on the Marsh. Albert took over the family farm, with the remaining brothers Harold, Henry and Bill. They lived for farming, seven days per week, 365 days per year, and thought of nothing else, which I dare say put a great strain on their lives in general, and consequently affected their physical and mental abilities.

Their farmyard had many traditional items. The old man's trap stood redundant under the shed, and they owned the only farm wagon in the village, which they used almost daily; it was a beautiful piece of carpentry. They also had a contrivance which, I believe, they called the 'mill race'. In fact, it was a piece of equipment very similar to the Claydon horse exerciser, which we see advertised these days. It was constructed in a circle, comprising a set of horizontal sails, which the horses were yoked to, and as they gyrated, the movement operated a system of cogs, shafts and gears, which empowered the barn machinery, i.e. winnowing machines, mills, straw choppers etc.

Despite being set in their ways, the Everatt brothers purchased their first tractor, a new John Deere, with similar excitement to a child on Christmas morning. Now, this really gave them a boost. The government, during the war, set up the 'War Agricultural Committee' and between them did a lease-lend deal with America to supply these affordable tractors to

*The Farms and the Farmers*

the British farmers, and so improve food production. They were mainly John Deere and Minneapolis Moline. The one the Everatts acquired was a three-wheeled John Deere. Having the row crop attachment, they were ideal for the inter-row potato and root crop cultivations, very easy to manoeuvre and a very great asset to their farm.

The Everatt brothers never took to operating the tractor very efficiently, and so they employed a younger man, George Lambert, who had already become competent with the tractors. Albert had inherited his father's shrewdness. He was a very good farmer, and could make money. They always had the yards and sheds full of large home-bred and Irish bullocks. Sometimes if they had some surplus fodder in the spring, and perhaps some potato pies (clamps) — if the trade for selling of the potatoes had been slow — he would buy in another batch of what he called 'half-finished bullocks' which would be anything up to 12–14 cwt. already. He would get them finished and away before there was time to clean the manure from the yards.

They were also good horsemen — too good, perhaps — for they would feed them the same rich food, when they were not working every day. Then they became difficult to handle — super fit, in other words. I remember watching them fork a large load of manure onto a cart with big Captain yoked in it, who had to be held by Harold, because he would not stand until it was loaded. Then, when it was time for him to go, he lunged and lurched, and pretended he could not pull it. Albert took off his flat cap and lashed him with it, which one might imagine helped the drama no end. However, in the end Captain set off at full speed ahead, while Harold hung onto the reins with all his might. Albert smiled and said to us lads, 'I wish I had not hit Captain now.'

Unfortunately, these very good people were not to be blessed with long living. William, who had not enjoyed the best of health, died at the age of forty-six in 1931. Harold was killed in an accident with the tractor and trailer he was riding upon during the early 1940s; and Albert, at the age of fifty-nine in 1947, was mysteriously struck by a goods train on the railway line which ran at the back of their paddock. He sustained serious injuries and as a result died very shortly afterwards. It was very sad, and so unfair that these good and so innocent people had to die so tragically, and especially so for Miss Ruth Hutton, who had been their loyal housekeeper and companion since their parents had passed on. Furthermore, it was to be the end of well over one hundred years of this family farming East End House Farm. Due to some family complications, and the event of Albert's sudden death, the estate had to be wound up; but I am pleased to say some descendents of these thoroughbred farmers are still thriving in the area today.

Mr George Lapish was their neighbour on Croft Farm, which he inherited from his father when he returned after several years working in Australia. He and his wife had no family; they worked their farm together with an additional horseman living in with them. George was a cantankerous character, but not in a nasty way. He was a chapel man, never smoked, swore, or drank alcohol. He was very intelligent and loved a conversation on any subject. He was also forward-thinking, but so was one of his colleagues, Mr George Leighton. As we were all having dinner in the field after a morning's potato picking, suffering from backache, leg ache, sore fingers etc., the conversation turned to machines, and George Leighton said there would one day be a machine to pick potatoes. Mr Lapish promptly chopped him down and said he was talking rubbish, as they would have to create one with a brain to do the job.

How they would marvel if only they could come back and see the Grimme Potato Harvester at work − in my book, one of the finest pieces of today's mechanical technology.

George was also an air raid warden during the war. Therefore, he was on call night and day to do his duty by touring the

*The Farms and the Farmers*

village, blowing his whistle and warning us to get prepared after he had been notified of an enemy air raid in the distance. It was a more reassuring sound when he was on his second tour, ringing the all-clear bell. Mrs Lapish was a very efficient and thrifty woman and a great asset to George. She baked bread, cakes and pies for sale at the door. They were of excellent quality and sold within an hour or so from baking. She also made butter and curds from their own fresh milk. They killed pigs so they always had a good supply of bacon, hams and lard for cooking. They also had a lovely large orchard with several varieties of plums, apples and pears, raspberries, gooseberries, red and blackcurrants, blackberries etc., and a large vegetable patch. All these fruits were carefully harvested and preserved, giving an all-year-round supply.

Mrs Lapish would ride her old cycle to Howden each Saturday afternoon and buy the same items every week: tea, sugar, yeast and salt. She used to say many a time, 'If we could grow these damn things it would save me some puff,' when she had been cycling against a headwind. The farmers' wives had their flour delivered in 10-stone sacks, saving a bob or two by doing so, and also ensuring a reliable supply. When we consider there were no refrigerators or deep freezers in those days, it denotes how skilled these people were in preserving and containing their food; and I maintain it tells us how near they were to being totally independent and self-sufficient, and not depending on anyone to survive. Tea and salt were the only items they took no part in producing.

George never owned a tractor, and when he retired and sold up, his successor allowed him to stay on in the house and help him on the farm for a week or two while he found another place. But once George got seated on Mr Harry Drury's new 'David Brown' tractor, it became more like a year or two; he was absolutely besotted with it! My elder brother worked for George for one year, so I spent a lot of time with them. He kept and bred some fine horses, and he used to break a young one in every year, and sell it during the following spring when the next one came of age. He would spend a lot of time with them to ensure they were perfect in all gears.[1] This helped to sell them, and he very rarely had any comebacks.

Being a dedicated horseman, George had no patience with cows. He kept three house cows and used to have some dramas with them. I recall one very hot summer's night when he was trying to milk them. The place was full of flies, and the cows were very irritable, their tails were swishing, and legs kicking, and he was cursing. The pails were being kicked over, and there was cow dung everywhere. Consequently, he ended up with only one small pail of milk, only to go on to a further calamity. He had cheered up and made the best of a bad job, and said to me, 'Never mind, my lad, you can take them back to the field now.'

He put the pail of milk down and picked up the broom to clean his boots before taking it to the kitchen, and knocked it over with the broom! So the final results of that episode were nil, and my brother (who was working for and living in with them at the time) told me the atmosphere was rather tense during supper that evening, although he knew nothing of the reason, for he had missed it all while he had been taking the horses to the far marsh to graze for the night.

Needless to say, we had a good chuckle when I enlightened him, but I must admit this was a one-off, for they generally had four or five gallons of milk yielded each day, which had to be dealt with immediately in the hot weather. The separating machine and the butter churn were hand-cranked, which was quite hard work for a woman. If any milk did go 'over the top', it was not wasted; the pigs soon made short work of it.

Mr Frank Heseltine came to Ashgrove Farm in the spring of 1944, following the Barkers. A client and good friend of

---

[1] Prior to a farm sale, on the sale bill or poster, each horse's credentials included in the sale would be stated, and one often saw the words 'Quiet in all gears'.

Frank's bought the farm on his behalf – 50 acres for £1,100. He put down a good deposit, and very quickly made the rest of the money, and became the owner of his farm within a year or two. He made no secret of the fact that he and his wife had five pounds between them after laying all their money out, to start farming, and he always said it was the first milk cheque that set them on their way to success. As he said, 'Being able to draw on a milk cheque one month after the day we started farming gave us a positive kick-off.' The next one soon followed, and they were away. They started with five cows, two horses, a few poultry, a plough, some harrows and a cart, and that was about it. Frank was a very fit, able and fully experienced man – absolutely determined to make a go of it. He ploughed all the money back into the farm, increased his cowherd into teens, and really put his back into cleaning up the land, which had not been farmed very well. In fact, some of his corn in the first year was so poor that he reaped and burned it, but this situation didn't last long. Within a year or two there wasn't a weed on the farm. Mrs Heseltine, too, was equally as keen. They did, in fact, work the farm between them for a few years, until their son was born.

Unfortunately, my parents left the village in 1946, which I was not very pleased about. But I kept in touch, and on leaving school in 1948 returned to start my working days in farm service with Mr and Mrs Heseltine. My first wage was £3 for a 47-hour week, half of which was deducted for board, with some extra for overtime. Frank had, by this time, really smartened the place up, extended the cow houses, built piggeries and drained some of the land. He was manuring the land heavily and buying wagons of shoddy manure. We were growing 20 tons of sugar beet per acre on 24" rows, and all the crops were above average.

The first year I was with him, he taught me his very sound, practical farming methods, and how to use a pair of horses, and do all other farm tasks. I remember planting six acres of potatoes by hand, and he set me off covering them over with the horses and plough, one row at once. This took several days, and he had the nerve to leave me in the field after a short briefing to finish covering each day, while he would go off to see to the cows! This was regarded as a man's job, quite different from ploughing. The wheels were taken off the plough to avoid moving or damaging the seed potatoes, thus leaving only the various hitching points on the plough's beam to control pitch, depth and so on, the onus being placed mainly on the man holding the plough handles.

Frank also trained his line horse to walk on the ridge top when covering potatoes, again to avoid damaging the planted seed with its large shod feet, as did many other horsemen and farmers. Frank was a very keen man and a perfectionist in everything he did. He would sometimes become a bit uptight if my horse work was not somewhere near his standard, and if the horses were playing me up. It always amazed me how those horses would jump to it when he entered the field, and they heard his positive voice. At the same time, I was lucky to have a man with a good sense of humour, and a lot of patience to teach me the basics of farming. While he could grow almost anything, his heart was in dairy cows, and unfortunately mine was not. I had, in fact, secretly made up my mind that I was destined for an all-arable farm.

As luck would have it, things worked out for us all in the end. My elder brother joined us at Ashgrove initially for a few weeks from Riseholme Farm Institute, where he had studied dairy farming. Theoretically, therefore, it was obvious to us all that he was the man for Frank. Hence, we had a problem. We were too thick on the ground, as they used to put it. However, at the same time, Mr George Ellwood, of Eel Hall Farm in the village, required an arable man, so the inevitable happened. Frank and Mr Ellwood had a discussion. I was keen to get to a larger arable farm we all knew each other well enough, so it was unanimously agreed that I took Mr Ellwood's vacancy, and my brother replaced me by staying on permanently with Frank.

This was 1951, and I was in my second job, a little sad perhaps to leave Mr and Mrs Heseltine. They had been very good to me and given me a start, and a very good idea what working for a living was all about. I soon settled in with the Ellwoods, who had a lot more land, and so I had the chance to learn arable farming on a wider scale.

Mr Ellwood was also a self-made man. He had started farming at Warren Farm, Barmby on the Marsh, in the early 1920s, and later moved to Eel Hall Farm, Asselby, in 1932, leaving Warren Farm, which Eric, one of his sons, occupied later. Mr and Mrs Ellwood had four sons and one daughter. It was a very sad loss for them when the eldest son, George, was killed in action in 1940 while on a bombing raid during the Second World War. He had done so well to become a pilot in the RAF, and they were so proud of him. John took a trade and became a joiner, and Percy and Eric took to farming.

Mr Ellwood was well into his sixties when I joined them, but he still took an active part in the farming. He was a jolly chap with a sense of humour similar to that of Ronnie Barker. He constantly had me in stitches with laughter, and always had some saucy tales to tell us, and how they had struggled on through the Depression, hoping for better times. He also said the rent was £2 per acre per year for Eel Hall Farm when he took it on. This is quite unbelievable when we compare rents of today to then. They had about 180 acres, with the two farms taken together.

When I joined them, they already had two Fordson 'Major' tractors. The team of horses previously mentioned had gone, but they still had three, which George referred to as 'ornaments'. They still had their place, though, as we used them in the carts at potato time, and they were still handy at harvest time when carting the sheaves, for moving the wagons, trailers and drays etc. from stook to stook; but the tractors hauled them home much faster than the horses could. Unfortunately, they were rapidly becoming redundant. They had another go at breeding during 1953. We put the best mare, a pure-bred Derby Shire, to a Percheron stallion that Mr Arthur Hammond brought to the village. 'Delevelle Limden' was his registered name and believe it or not, we all got a surprise, for the mare foaled twins. Now, this was a novelty! Here was Mr George with the first set of twins he had ever seen, never mind owned, in his seventy years of life. Even though horses were on the decline, we had people from far and wide looking and photographing them.

As Goole's Agricultural Show was approaching, Mr Tom Clegg, the auctioneer and commentator of the Goole Show, came to Eel Hall Farm and asked Mr Ellwood to enter them, which we did. They were the centre of attraction and we won a special prize for the foals, and I must admit, we had very little competition. Needless to say, these foals became a liability, and were sold unbroken to a dealer at two years old. Mr Jim Nut, of Gilberdyke, also travelled a Shire stallion up until 1953 -- 'Haxley Lad' was his name. I believe these two heavy stallions were the last to enter Asselby village. I was taught to use a tractor by Percy, Mr Ellwood's son and the head man on the farm. He was a first-class farm man and an excellent hand at making corn stacks and thatching them, which again, he constantly tried to get me into, in spite of it being a dying art with combine harvesters just around the corner.

A lot of farmers, when the tractors came on the scene, adapted their horse implements and machines for use with tractors, although this idea was not entirely successful in some cases. Take the mowing machine, for example, which had three operator's levers, one to lift the cutter-bar up and down, one for in and out of gear, and the other for cutter-bar pitch, which were easily operated while one sat on the machine's seat with a pair of horses hauling it. When using the same machine adapted for a tractor, the driver could not reach the levers, so it was either one man on the reaper, and one on the tractor, or the tractor driver on his own, climbing on and off the tractor at every end of the field or corner, as it was a must to lift the cutter-bar clear of the cut material when turning; and believe me, this was a very tiresome task.

One day, Percy and myself went to cut a six-acre field of clover with the tractor and an old adapted reaper. It was going badly: the knife kept stopping and blocking up the points. We tried everything we knew to make it go – all to no avail. Alas, we both expected harsh words when we returned back to the yard at twelve o'clock for dinner.

'Have you got it knocked down?' the old man exclaimed.

'No,' replied Percy, 'we haven't cut half of it yet.'

'Strewth!' the old man growled. 'Two men, half a day, three acres! Many times I have cut more on my own with the horses.'

When he had finished, Percy replied, 'It's time we had a new reaper. The damned old thing is worn out. You only give five shillings for it twenty years ago — what more do you expect from it?'

'How do you know what I paid for it?' his father snapped.

'Well, you have boasted about it often enough,' said Percy. And the saga ended with us all in a fit of laughter.

'Well, I suppose you are right,' said George. 'Go and have your dinners, and tell your mother to keep mine while I return.'

Then he reversed his Morris 10 motor car out of the shed, and went to Glews of Howden and bought a new 'Bamford Royal' model, which we took delivery of later that day. It was, of course, a model built for use with the tractor, and we greatly appreciated it.

To get back to George's motor car, this Morris 10 was a very strongly built vehicle, with not a tiny bit of rust on it, that he had acquired before the last war; but his driving skills held nothing to be desired, I must add. He was not the only one like this by any means, but unfortunately the older he got the worse he became. In fact, Mrs Ellwood and their daughter, Grace, refused to ride with him in his later years. The sons did not have to, as they always drove when they went anywhere together. I must admit he gave me a few scares, but I had plenty of confidence in him. Sometimes, when he had to take me from farm to farm, the family would make fun out of it, and say things to me like, 'What kind of flowers do you prefer?' or 'Are your insurance policies paid up?' but the old chap took it all in good part, and I think it goes without saying that one cannot compare anyone's driving standards with those of today.

Mrs Ellwood, too, was a jolly woman; she was very punctual and she expected others to be also, particularly at meal times. I can honestly say, I never once waited for a meal in all the seven years I lived with them, and it was a crime to be late at any meal time, but the poor lady suffered terribly with arthritis in her later years. I once laid myself wide open for her to tick me off in her stern but joking manner. We had killed a pig, and were having spare ribs for dinner, which were delicious, with all the trimmings, vegetables, potatoes, etc. The farm ladies really put some top-class meals on the table, and there was always plenty in case any extra people should turn up unexpectedly. The farm dogs and cats, too, lived on what came from the kitchens.

On this particular day, Mrs Ellwood asked me to take what was left of the spare ribs to the dog down the yard. As I walked down the yard, I thought, Fancy giving a dog this! and could not resist having a bite at the ribs. Of course, she had seen me through the windows, and when we had tea, she waited until we were all finished then asked me if I had had enough. When I said, 'Yes, thank you,' she replied, 'I do hope so; then you won't have to eat the dog's meal, will you?' Of course, it ended in laughter and me feeling very embarrassed indeed.

They did not have a mill race at Eel Hall, at least not as long

*The Farms and the Farmers*

as I have known it. The system for barn machinery was powered by a drive wheel which was driven by a portable engine, by hiring the local threshing contractors, or by steam engine for a required time. The drive shaft and wheel are, in fact, still intact on the barn wall at Eel Hall Farm to this day. I can also recall the last time it was used. It was 1941, and Mr Ellwood had two bays of straw left over from the previous harvest, which was in the way of the new corn due to be stacked in there. They were waiting for harvest time and had very little to do, so they chopped the straw and carted it onto a fallow field. I suppose one could say this was medieval straw incorporation. Incidentally, they powered the chopper with the standard Fordson tractor.

George was the son of Samuel Ellwood of Elm Hurst, Asselby. He was the village joiner and wheelwright and had five sons and one daughter. They were all highly skilled and successful people. Alfred and Bill farmed away from Asselby. Tom was the blacksmith. Fred took the old man's trade and became the second village joiner, and Alice married William Everatt, another village farmer. George had done very well for himself and bought both his farms. In fact, he bought Eel Hall during the mid-1950s, paying £14,000 for 70 acres. He retired shortly after.

Farming was changing rapidly, and I became restless and desired a post on a larger establishment with more to offer. So I left the Ellwoods in 1958 and moved into Lincolnshire, but one can never forget such fine and genuine people, and I was also the last man to be in farm service in Asselby. 1958 was a very wet season and this was another reason, I think, why I became bored, fed up and ready to move on. We were doing nothing weeks on end due to the rain. Had it been today, I suspect we would have been stood off, but there was none of that in those days. The farmers were loyal and honoured their agreement with us, but it was a very frustrating time for all concerned. When I left at August Bank Holiday time, the farm looked a mess. It had never been dry enough to earth up the potatoes, the sugar beet and other root crops were lost in rubbish, and the clover and hay lay ruined in the fields. They had cut some winter barley on the farms, which I moved to with two Massey Harris combines. These were only small machines compared with today's giants, but they had made ruts 12"–14" deep. The first job I was given there was to spin three straw trails into one. They were very thin due to very difficult operating conditions. 1954 was also a very wet harvest time. It was, in fact, the only time in my life that I have seen grain sprouting in the ear of standing corn, because it was ripe, and had not been cut due to continuous rain, resulting in some very poor samples of grain.

Mr William Gardam Everatt farmed Box Tree Farm; he had no relation to the aforementioned, but was very similar in style. He was very tidy minded; his stack yard looked a picture of art after harvest with the corn ricks built as plumb and square as a house, and so neatly thatched, it almost seemed a pity to disturb them when the thrashing machine came in. Everything else in the farmyard, too, had its place, and their stable was immaculate as well as its contents, particularly the horses. They only kept four, but they were the best, and the pride and joy of his two horsemen, Harry and Ira Coatsworth.

It was a splendid sight. We would sometimes stand and watch them bring them out of the stable on a lovely spring morning. They were immaculately groomed and elegantly geared up. Mr Bill would help them pair them up, as they were so fit and eager to be off to the fields. He would stand and admire them, and if he was satisfied with their appearance, he would give the men a leg up onto the horses' backs, and off they would lunge at a half-galloping pace. There would also be other farmers on their way, too, which filled the streets with that familiar clip-clopping of their large shod feet and the jingling of trace chains, the blowing of nostrils and the aroma of them — a truly magnificent sight.

*The Farms and the Farmers*

Harry was an excellent horseman. He lived for them, and spent all his spare time cleaning and caring for them. He wanted to have the best in the district. Mr Everatt never owned a tractor and Harry and Ira were just not interested in them at all, and in fact never ever drove one; but they did hire contractors in for deep ploughing etc. Harry often told me in his later years of a time when he and his brother were binding a very heavy crop of wheat on the river-side land. It was very hot weather, and after two hours' hauling the binder the horses became tired and lost their essential pace, thus making the task harder. Dripping with sweat and plagued by flies, they were giving them a rest when Bill turned up with the afternoon tea.

'Are you trying to kill them?' were his first remarks, referring to the horses, of course.

'No, we are trying to get your wheat cut,' replied Harry.

They sat down and had their tea and Harry noticed Mr Bill was in deep thought. When they arose he told them to unyoke the horses, put them in the bait shed, stook up the wheat they had cut then bring them home. Most of the farms had a bait shed at the far end of the farmland, which was a considerable distance from the homestead, and they used to feed the horses in these sheds at midday when they were making one yoke,[2] and at potato harvesting time etc. It also ensured they did not catch a chill while eating their feed, as they were often in a sweaty condition.

Bill not having enlightened them of his plans, Harry became anxious and dwelled strongly on the possibility he might be contemplating buying a tractor, and could not resist calling after Bill as he mounted his cycle, to enquire if this was so — only to receive a wry smile. As they stooked up the wheat, and tried to envisage what Mr Bill had in mind, they

---

[2] A one yoke was the best way to achieve maximum output from a twelve-hour day, which was a horseman's usual working day. Normally, they would return to the farm at noon for an hour's break but because most farms had a far field, which could be anything up to a mile or so from the farm, it was not viable for horsemen to do so. Hence the one yoke and the short break were often taken in the field.

noticed another cycle coming towards the field, and they soon realised it was the village blacksmith with his tool bag. As he entered the field, he chuckled and said, 'Now then, lads, Bill says you have a binder which needs operating on.'

He had come to modify it to be hauled by a tractor, which was a simple task. All it required was the pole shortening, and three holes drilling to fasten the iron plates which connect the binder to the tractor.

With this done, Harry knew his beloved horses would never have to drag the binder again, which gave him some consolation, but he dared not ask the blacksmith if he knew anything further, for he knew the blacksmith would have really wound him up, as he was full of that kind of fun. So he had to wait while they came back home. When Mr Everatt told him he had gone to see Fred Masham, the local contractor of Barmby on the Marsh, Harry was delighted. Mr Masham, a forward-thinking man, had by then established himself with a fleet of tractors, and was in constant demand. He would hire out the wheel tractors and a driver to haul the farmers' machines. He also had two crawler tractors for deep ploughing, two lorries for carting sugar beet and cattle etc., and very full order books.

Farming methods were changing fast, and as Bill was getting on in years, and had no desire to move with the times and no family to succeed him, he retired, and let his farm to the Arminson family. One of the conditions was that they did not disturb the horse stable or utilise it for any other purpose. This should tell us all what a true horseman he really was.

I can clearly recall the unexpected death of Mr Cyril Bolden, of West End Farm, who died suddenly at the Great Yorkshire Show of 1957. He had become a little tired in his later years, and while making a special effort in scrambling up a bank to get a better glimpse of the Queen, he was struck down with a fatal heart attack. This was a very sad loss to his family and friends. He had struggled along on his small farm through the Depression years, and brought up his three sons and one daughter, who all stuck to farming, and today the younger generation of Cyril's family are a credit to their late grandfather. He was a man with a sunny personality, no matter what dilemma he was confronted with.

Being a keen and competent league darts player, Cyril played in the Howden and District Darts League, and so did I. We played for the King's Head of Barmby. He looked forward to the matches with great enthusiasm and, again, his friendly down-to-earth personality made him a very popular figure amongst the players whenever we were at away matches. Towards the end of the season, we were playing two or three nights per week. As well as the league matches there were the pairs and individual league and brewery cups to compete in, and lay-off friendly matches and so on. Cyril was a lob thrower, which at times created much laughter — and also anxiety, for he displayed a 'couldn't care less' attitude when other players would be serious, if not tensed up.

During a dead silent important match, or any other event, this appearance proved to be false, for he could be relied upon to finish when it was up to him to do so. Often his darts were virtually horizontally hung in the board, but he could usually put them where he wanted. This was very often three in double top which, of course, was double twenty, which all the best players went for, for obvious reasons, as one never saw a treble board in that area.

The season would end with an annual darts dinner and awards presentation in the Shire Hall at Howden. Such a grand evening, to be amongst twenty or so teams of smartly dressed darts players and their ladies and followers, is unfortunately something I have not seen or been involved in ever since. Cyril's farming took a turn for the better when, after the outbreak of the Second World War, the War Agricultural Committee advised and assisted these small farmers to take up milk production. He was keen to do so and went into it with zest and zeal. He was a good feeder and produced some top quality milk which, I believe, was at a premium and he was justly proud of it. He was also amongst the first to venture into buying a tractor and bought a Ford Ferguson 1926 model from Mr Askham, of Newsholme, with iron-cleated wheels and a wooden steering wheel. He soon had a two-furrowed plough and an old cultivator to use with it. The elation showed in his face as he crawled up and down his fields at a pace perhaps no faster than that of his horses, singing happily, as singing was another of his great loves.

There was no doubt in Cyril's mind that it was money well spent. For one reason, most of his land was quite strong clay and difficult to plough with horses and, as he often stated with a chuckle, 'She' — referring to his tractor — 'never needs a rest, never needs to stop to pee, and never wants to stop for dinner!' This was so typical of him.

Mrs Bolden, too, was a lovely lady. In my mind, she made the best scones for miles around. She used to bring out a large basketful of them on threshing days, topped with that delicious home-made butter.

The threshing days were busy ones for the farm ladies, as the day or half-day minimum hire contract for the threshing set included meals for the two men operating the set. The engine man would be on the farm by 6 a.m. at the latest to prepare the engine and get up the steam ready for a flying start. His mate, known as the feeder, joined him later, and then they would enter the farm's kitchen for breakfast, and perhaps discuss various aspects of the day's threshing ahead of them. The farmer also supplied the rest of the team that was required to thresh, a minimum of ten hands, with tea and food twice a day. Sometimes if they had a long move, e.g. from Asselby to Spaldington, they would politely tell the lady of the house not to prepare an evening meal for them, in order not to waste any time, and get the set on the road, and set up at their next destination ready for another 6 a.m. start.

*The Farms and the Farmers*

The Palmers, their neighbours, farmed Crossings Farm, and three generations of the family lived happily together there for a good number of years. One often found this in the farming fraternity, particularly in a case of an only daughter or son, but having said this, it does not say much for the adage so often heard, 'There is no house large enough for two ladies.' However, Mr and Mrs Arthur Palmer's only child and son, John, married Miss Doris Hutton of Asselby, and they and their two children lived together until the old people died and John took over the farm.

The grandfather was the steady type who, in his later years, was prepared to stand or sit back and let them get on with it; but Grandma was the exact opposite. She was up at the crack of dawn, the washing done and hung out by 7 a.m. and several cows milked by hand. She also kept a lot of free-range laying hens, which she would not allow anyone else to tend. She also helped in the fields. As well as being a totally self-motivated person, she was a happy, wise, thrifty and formidable old lady. She once took the afternoon tea into the harvest field, which John was 'opening out' at the time. This entailed mowing with a scythe the perimeter of the corn field and tying it up into sheaves to make a track for the binder. Then it could make its first run without running down and losing any corn. The old lady asked John what he was about to do next, as he had told her he had finished that field. Then she noticed a run of mowed corn, which he had not tied up, and exclaimed, 'No you have not! What about this?'

John bit hard on his pipe, and his face beamed, then it turned into his characteristic belly laugh as he replied, 'That can stay there, there are too many soldiers in it for my liking.'

'Rubbish!' the old lady replied, and promptly set about tying up the corn he had left with her bare hands.

The 'soldiers' which John referred to, of course, were thistles, and that was their nickname. They were well known by anyone who had worked amongst the corn harvest, as they were perilous in the sheaves of corn, especially when hand-handling them. I often think it was these trivial 'waste not, want not' values these older people held that kept them afloat, and the saying I heard later in my farming career, that the larger farmer wasted more than the smaller one grew, was perhaps not a fallacy. I believe the strategy of the farmer, which was to produce a little of everything, enabled them to always have something to sell and the Palmers excelled at this.

John was six foot plus, a very handsome and athletic chap, smoked twist tobacco in his pipe, had a very jovial personality and had the knack of being able to turn any serious predicament into an hilarious drama; but woe betide anyone who dared to deliberately upset or double-cross him in any way, for he would turn like a raging bull. This was sometimes highlighted on the cricket field, perhaps not the appropriate place; but he was the 'Fred Trueman' of the Asselby team for many years, a very good all-rounder, in fact.

He did not like a stonewall batsman and soon made it obvious to them. He was also a keen ice skater, but his main hobby was game shooting which he absolutely excelled in. This, however, was an entirely different kettle of fish to the strictly organised and, in some cases, segregated shoots and syndicates of the present day.

The interest declined between the wars amongst the shooting fraternity and consequently it lapsed, in some cases into pot hunting. Very few of the farmers were interested, and they did not pay much attention to people straying on their land with a gun − within reason, of course, which suited John. Although the word of mouth agreement which he had with most of the other farmers was beneficial to both parties, for one reason John was an excellent man on threshing days. He could handle those sixteen- and eighteen-stone sacks of grain with ease, for days on end, and was the best man on a corn stack for miles around. Those were the only jobs he ever did on threshing days. While most of the threshing was, as regards

*The Farms and the Farmers*

men, worked on a borrow and payback basis, again this was beneficial to John, for he would perhaps have only one day's threshing per season, and then for eight or nine days, he was on pay; and by permitting him to walk through their root crops in the shooting season they kept in each other's good books.

So, John, Harry Coatsworth, Dick Clayton and John and Eric Ellwood were Asselby's current shooting gang. It was nothing very formal, no beaters, no reared game, just them and their dogs. It was good English and French partridge country, and hares and rabbits, with a relatively good habitat of potatoes, turnips, sugar beet and kale fields. In these they walked up with the dogs hunting, hopefully not too far ahead of them, and followed the coveys around, sometimes setting others up while doing so. They were not always together, but some of them would shoot most Saturday afternoons. They all had very good dogs which were, of course, a very good asset to them.

John was the most dedicated shooting man. After a morning's harvesting, and on perhaps the last cricket match of the season, he would pop off on his own for a couple of hours' shooting before darkness fell. I shot with him much later on when I started work and acquired my first gun. He was by then a very highly skilled marksman but he would go to great lengths to train and pass on all the dos and don'ts pertaining to shooting. He was still very fast, and could have two partridges down from a rising covey before I saw them. Other times he would wait as long as he dared, and give us learners a chance before taking the birds, which otherwise would have been ours. He would then chuckle, and I won't write down some of his comments, but he really was a fine and outstanding man to be with. He had such a great wealth of knowledge on the subject and he always had time for people, particularly us lads so junior to him. He was also out most nights during the winter duck and goose-shooting season. The marsh ground of

Asselby and Barmby, before it was drained and arable farmed, was a haven for wildfowl; but, alas, things were changing. John had much more time to shoot after acquiring his new tractor.

New faces were appearing in the village. It seemed the old communal and flexible image was in decline. For instance, John and his company had always been allowed on Box Tree Farm land until Mr Bill Everatt retired and a new tenant came who laid in wait and took his chance to confront John and a few more who were walking up his sugar beet for the purpose of shooting partridge. A very nasty argument took place, and John became very irate, but he realised he might, after all, have to compromise. Then he thought, The hell I will! So they parted with bad feeling. This did not deter him in the least, for he chuckled as he told me of the drama, and I suspected that he envisaged that this was the start of a change, largely I think because we were all becoming better off.

Organised shooting was coming back into fashion. John had already obtained some shooting rights away from the village and was mixing with people in his category − Joe Wheater, for example, of Hull, the famous pigeon shooter. This was another of his favourites. In the spring he would spend days on end shooting wood pigeons on the spring-drilled pea and bean fields, at the better wooded areas around Howden. It is no exaggeration to say he brought vanloads of the birds to Asselby.

I cannot pretend to be much more than ordinary with a gun myself because I was never keen enough. John knew this, although he never said so, and never stopped encouraging others. I still own a gun and all the paraphernalia one requires to legally do so, and I suppose it goes without saying every time I use it, yes, that extrovert man, John Palmer, flashes through my mind.

Incidentally, he was the first person to buy a television set in Asselby. About 1950, he purchased this wonder machine, which was the talk of the village. In spite of it being a 10" midget screen, fraught with problems, i.e. poor reception, a lined picture, interference from other stations and the like, John was understandably proud and quite undeterred. He constantly persevered with the controls to obtain a decent picture, and the first evening he invited me into their home to view the thing, his mother made it quite clear that she thought it was a complete waste of money, and a hindrance to life in general. The rest of the family were in bed by 9.30 p.m. and Grandma sharply reminded John it was time to switch off, as she too went to bed. He was elated to be able to view the test cricket live, which created much antagonism between them as it was in work hours. However, as mentioned before, John tolerated no nonsense from anyone, particularly a woman; so life went on, and it soon became obvious television was here to stay.

Mr Fred Johnson married Mr and Mrs Brabbs' daughter, of Village Farm, and later took the farm over. He also took on Field House Farm, which belonged to the estate after Joe Howden went bankrupt on it. Mr Alf Andrew was his foreman and he occupied Field House farmhouse. There is one thing I am quite certain of: he would not have found a better man anywhere. Alf and his family were a great asset to Mr Johnson. Alf, his two elder sons Frank and Alan, daughter Margaret and Fred's son, Albert, worked the farms. My younger brother and I spent a fair amount of time in Field House farmyard, as we were regular buddies of Alf's younger sons, Arthur and Geoffrey. Our favourite treat was being allowed to help with the feeding of the livestock on Sunday afternoons, which the men were allowed to do early on a Sunday in order to get a few hours off. So, at around two o'clock they would start the system with the horses, which were tied in the stalls, turn them loose into the fold yard for a drink, and a walk around while one cleaned out the stalls, put in a feed and added clean bedding straw, and one large mangold in the manger. This I shall never forget; I was six years old when I fell for the task of

*The Farms and the Farmers*

putting the mangold in big Boxer's manger, and before I could leave the stall, Arthur let him come in. Now, he was a monster, and I was none too happy at the sight of him eagerly rushing back into his stall, to get to his food – which, of course, was all he was interested in, not the fact that I was also in there. We soon became used to them, though, and would sometimes enter the stable and have a go at brushing them, and sitting on the older one's backs, but taking very great care not to be caught in there.

We were very puzzled on entering the stable one day. All the horses were out at their work, and there was just one in a stall, but the wrong way round, tied by two reins to an open bridle, and to each stall post. Arthur soon enlightened us that he was in the pillar reins. This was a young horse in training. I later learned this manoeuvre was to familiarise him to the bit or bridle, which was to be the main control point for him in the future. Whilst in the stall that way, the horse could not go backwards because of the manger. It was tied so it could take a short step forward and so put pressure on its mouth by doing so, and the same for whichever side movement it made. After several sessions of this, he was taken out in long reins, with a pretty good idea what all this tugging and pulling at his mouth was all about. It also got them used to being approached whilst in a stall, and it was very important that they were trained to expect to be approached from their right side at all times. This also applied to cows in stalls, which were much more stubborn beasts. I have heard of several serious accidents resulting from this rule not being adhered to.

To get back to our Sunday afternoon, there were many other jobs we were allowed to have a go at. They always had thirty plus really large Irish bullocks in the fold yard and all the loose boxes full of younger stock and sucklers, cows and calves. Part of their diet was pulped turnips and mangolds. Alf would start up his Amanco engine, which powered the pulping machine, and we would throw in the turnips etc. as fast as we

## The Farms and the Farmers

could, hoping to block the thing up — but no chance, we tired first. It would probably take half an hour or so to pulp and mix the turnips etc. with sugar beet pulp, chaff, ground cereals and other additives in preparation for the next morning feed. Then we would go over to the brick barn where the sweet-smelling oat straw was stored away from the winter weather, as it was good fodder. One of us climbed the ladder onto the stack and threw twenty or so bats[3] down, while the rest of us carried them to the fold yard, and the bullocks would start to eat it immediately. Also in the barn there was a large stack of cotton and linseed cake slabs, which they also put through a breaker before feeding to the cattle.

Field House Farmstead was one of the best laid-out farms in the village, in my book. The architect had allowed plenty of room for all aspects of farm work. The large stockyard on one side bordered the brick barn, the stable and the fold yard, and diagonally opposite stood the wagon shed with granary above. Therefore, on threshing days, the products were near their destination, and the buildings sheltered the stacks from the prevailing winds. The large fold yard had double doors at each end, allowing a through road to the grass field at the back and the fold sheds, backed to the north, provided shelter for the stock. The steam house and pig range were also sited together, again with plenty of frontage room for all manoeuvres, and traditionally nearer to the farmhouse, from where all by-products from the dairy and household, often referred to as swill, went to the pigs.

The farmhouse stood well away from the farm buildings, separated by the apple orchard and a soft fruit garden on one side, and vegetable garden on the other side. Also attached to the house there was stabling for the hunting and trap horses and horse-drawn vehicles. I cannot pretend to have paid any attention to these details whilst playing around that particular

---

[3] Bats are tied bundles of straw, which the threshing machines produced.

farmyard as a child but now, with hindsight, I see a brilliant example of architectural planning, perhaps relating to the high farming years of the 1850s, the golden age of Victorian agriculture.

There are many reasons why men such as Alf were at a premium. They were so loyal and worked seven days per week through the winter when all the livestock were inside. There were no days off, no relief men, and they were very rarely sick; they could not afford to be, because the livestock were absolutely dependent on them.

Field House was Mr Johnson's main farming point. He was not a regular working farmer but he kept a small herd of breeding Beef Shorthorns at Village Farm, which he looked after himself. He toured the farms daily on an old cycle and would sometimes help the men at very busy times. The business side, he operated from the kitchen. It was a sad loss to him when his wife and daughter, Hilda, died within a very short time of each other. Their deaths, from a mystery illness, left him and son, Albert, beside themselves, and he later employed a housekeeper, a Mrs Horsman, who stayed loyal to them for many years.

Albert became the tractor driver for his father as their farming progressed, which made life a bit easier for Alf and his sons. Mr Fred was a very good practical farmer. They would have up to 30 acres of potatoes to lift and pie in the autumn, which found us lads plenty of days helping to lift them. Three shillings per day was the going rate for hand picking the potatoes into lines of baskets along the field. These the men emptied into the carts and afterwards took them to the place in the field, where they were to be stored in a pie (clamp). The best lad was sometimes given the job of taking the loaded carts to the pie and returning with the empty ones. This was an easy number, and therefore there was plenty of competition for it. They always sited the pies or clamps, as they were sometimes called, with the future in mind, i.e. preferably alongside a farm road or on a high and dry part of the field, not too hard ground, as the pies were covered with 12"–14" of soil, hand dug with spades and shovels etc.

With his wife and daughter gone, life for Mr Fred became unsettled and to the peril of his health, he took to heavier drinking. So every evening he would don a smart suit, crank up his Morris 10 motor car and head for Howden town. Not a very exciting venue, one might assume, but night life in the small town during that era was quite lively, which we will come to later; and I must add this was during the Second World War. Petrol was on ration, and whilst the farmers were allocated extra petrol coupons for journeys relating to their farming business, the war agricultural officers and police officers had the power to stop and check any motorist who they might suspect of making a constant repetitive unwarranted journey. So Fred was one jump ahead of them. He once told my father he carried the same horse collar in his car boot for two years, and had he been questioned, this was his alibi: he was taking the collar to be repaired at the saddler's shop in Howden. The saddler, incidentally, was Mr Fred Soames.

Fred was, at this time, not an old man but, alas, not a young one; so this unaccustomed lifestyle he had adopted, in time, began to take its toll. He became mentally and physically slower, and took to chugging around the farms in his car instead of on his cycle. This was sad, particularly to those nearest to him and we had always found him a cheery, kind-hearted and genial man, full of playful badinage.

His state of mind was perhaps highlighted when during 1948 East End House Farm came on the market. He attended the sale, perhaps suffering with a slight 'wine headache', a drink-related mental condition familiar to all auctioneers, if not the beholder. Mr Fred's bid stood at £6,000 for three of the auctioneer's taps and, consequently, he became the new owner of East End House Farm, and I later learned that he was the most surprised, considering he had bought 90 acres of grade one land, a big farmhouse with a relatively large range of farm buildings, a king-sized stack yard, and a real lot of manoeuvring room. For that kind of money, it was a bargain in anybody's book.

The Everatts had farmed much more land at East End House Farm, which was rented land and belonged to the estate so, therefore, not included in the sale. So, Fred had visited a public auction and done something we all have done at one time or another − bought something he did not particularly want, or need! Hence a mutual discussion between them. Albert maintained they had enough land in hand to efficiently manage and, no doubt, he envisaged it would, in the near future, place more responsibility on him if they took another farm on board. Alf was prepared to go along with any decision Fred and his son, Albert, made on the issue, which was, in fact, to give up the rented Field House Farm. So Alf and his family later moved to the new establishment, and they became owner and occupiers of all the land they farmed. Incredibly, Mr Fred showed very little interest in their re-arranged farming policy or, indeed, scarcely anything else; and his health was obviously deteriorating. He was then involved in a slight accident when driving his car home from an evening out at Goole. No one was seriously hurt, but it left him very shaken. He later suffered a few periods of illness, which kept him indoors; but alas, finally during 1950, pneumonia confined him to his bed, where the poor chap peacefully passed away.

Again, Albert, Mr Alf and his family were dealt another sad blow, for this time it was their respected chief, the head of their little empire − the main man, so to speak. Albert was thrown in at the deep end in a sense, for he had had no training or involvement at all in the business side of their farming, and they had not discussed the future in his father's last years. However, there was some consolation. He had a good crew and a first-class captain, so between them they confidently sailed. However, believe it or not, there was yet another very sad time for them. Mr Alf fell ill with a serious illness, and sadly he too died in 1955. I feel no words I write here will suffice to describe how distraught, sad and utterly lost they all were after Mr Alf's death. However, life had to go on; but I am sure none of them got much consolation from that adage.

Albert fared badly at his role as master. He could not delegate, and he once told me he was no longer interested in making money, as he was in such a high tax bracket, and he also felt that no one could replace Alf. He played at it for a decade or so, but it has to be said he was no farmer, although he was a very good-natured, fun-loving, outgoing and sometimes serious chap. He adored dancing, and would often take a carload of us younger colleagues to dances as far as Bridlington on a Saturday evening, and we never missed the local ones. Therefore, I dare say it follows he was a ladies' man. This, according to some of his friends and self-appointed guardians, was what he lacked: a good lady to share his life, care for him, and motivate him in office. But it never happened. Albert pursued his bachelor lifestyle, and finally sold up his farms, and retired early, which was a pity really, but he felt the best days had gone. He often recalled the times when they were all together and how they worked together in such peaceful harmony.

Harvest time, for example, always seemed to be a time of glorious sunny weather when they were carting in the sheaves of corn. They were well enough aware of the urgency of the job but it was such a laid-back, stress-free, almost casual time, that there was always time to sit down to afternoon tea, and those lovely large trees of hazel pears had fruit ready to eat during this time, and us lads would pull at the lower ones. Then Mr Fred would arrive under the tree with a large piece of wood, fire it into the top of the tree and shout the order, 'Stand clear!'

*The Farms and the Farmers*

Then we had to wait until his pockets were full. He would then knock us a shoal of pears down and mutter something like, 'Don't go far away from the toilet, you lads, when you've filled your bellies with those.'

One time the piece of wood stayed up in the tree, and all attempts to dislodge it failed. So Mr Fred called the men making the stack, and asked them to bring the longest ladder, as he dare not leave it up there to come down whenever. When the drama was over, he playfully tried to blame me and said, 'Now don't throw that piece of wood up there again, my lad.' He then mounted his cycle and pedalled off back to the cornfield.

The master often took the role of forking the sheaves at the field end during harvest, sometimes to pace the job. Also, if there was a showery spell or heavy dew in the mornings, he would be there to make the yes or no decision to cart or not.

Albert obviously dwelled on these happy and stress-free times and he had no desire to adapt himself to the fast-changing world, and perhaps having the resources motivated him in that direction.

Mr Fred Harrison Senior and his wife and family farmed Back Lane Farm. This was a compact little farm with all the 70 acres of land in one block, previously farmed by the Farrahs, and it was said that it belonged to the Leaconfield Estate at one time. Of course, that estate bordered Asselby, and I suspect the railway, when it was constructed, cut through Leaconfield's bordering field, leaving some land on the Asselby side of the railway with no access. Hence the bridge over the drainage board's dyke, and the road leading to this cut-off land from Asselby, opposite School House.

Mr Harrison Senior was a large, jovial man, blessed with a good wife and family who helped run the farm. He loved a good laugh and used to come further into the village on a summer's evening when his work was done and sit on the corner stone by the chapel. These large stones were placed at the T-junctions of the narrow lanes in the villages to protect the boundary walls, and properties adjacent to them, which were vulnerable to the wagon and cartwheels. Should a false manoeuvre occur, the iron-rimmed wheels struck the stone. This was tapered, and therefore created a slewing action to the horse-drawn vehicles, eliminating any possible damage. The one Mr Fred used as a seat is still intact. He would encourage a gang of us lads to get up to all sorts of tricks, such as setting us off racing on various routes around the village on our old cycles, or running. With his large pocket watch drawn and held in hand, he would time us and set handicaps, and cause squabbles before handing over the two or three pence he had put up as prize money.

Of course, I was very young at that time, and can just remember his youngest daughter, Kathleen, coming to our house, as she and my sisters were friends, to inform us that her father had had a heart attack and died. Now, that kind of news to anyone so young does not sink in for some time, and my four-year-old brother, Gerald, and I were not convinced until we went down the lane the next day, as we had so often done, and he was not around in their farmyard. We missed him immediately because he had two large white Wyandotte cockerels, which he bought at Selby market, called Jim and Jerry. These cockerels would come charging at us lads in a menacing mood, and Mr Harrison made us believe he had them at his command when he drove them away with his stick. Other times he would spot us coming down the lane and take up a discreet position and watch us warily pass his yard, and if Jim or Jerry were about, the sight of us legging it in sheer terror reduced him to hysterical laughter; but usually, if he saw us hesitating, he would reassure us and encourage us not to show fear to these birds. Being such a kindly, cheerful, fun-loving, and not the slightest bit indifferent man, it was no wonder it took us a while to realise he was no longer going to be there. Obviously, it was much worse for his family, but fortunately they were very capable of carrying on. So Fred

Junior became the master, and took over the arable side of the farm, while his mother and sisters did the yard work.

Young Fred had inherited his father's personality, and was so much like him in many ways, although being tall and slim he featured his mother in that respect. I worked quite a lot with him during harvest and potato picking time; the Ellwoods and he would work together, as quite a lot of the farmers did in those days. It made for a better team, at potato time especially. A good team of hand pickers would command four or five carts, two horse teams ploughing out and harrowing up, etc., and plenty of men to keep the pies covered, baskets emptied and so on.

Fred held some quaint beliefs. One of them was, 'There are tricks in all trades but the potato trade is daylight robbery.' As is the case today, supply controlled demand, and he used to have some heated arguments with the merchants. He once grew some Arran Banners, a round potato which tended to be large and hollow if grown on rich land. Charlie Batty, a merchant from Hull, took a sample and later agreed to move them. Five, sometimes six tons was the current load. They were sorted into 1 cwt. sacks and lifted onto the lorry by hand. The lorry arrived for the first load, and when half the sacks of potatoes were lifted onto the lorry, a sack came loose and revealed the contents. The actual words which were exchanged between Fred and the lorry driver I dare not disclose, but the gist was something like this:

The driver, as he picked up the spilled potatoes, said, 'They are a poor sample!'

Fred: 'That's none of your business. Let us put the remaining sacks of potatoes on, then you can be on your way.'

The driver cut a potato in half and found it to be hollow. 'Sorry, boss, I cannot take them.'

Fred: 'How many times do I have to tell you? It is not up to you, so let's have them on, and you get off to Hull with them.'

'Sorry, boss, I cannot take them — and that is final!'

Fred: 'But your boss has been here and seen the potatoes!'

Needless to say, the lorry went away empty, and a feasible explanation for the saga was a well-informed and brainwashed lorry driver, and a sudden glut of potatoes on the markets, giving the merchants 99 per cent of the bargaining power.

Fred also recalled the time when it was vice versa. There was a spell of hard frosty weather, which caused the farmers and growers to be reluctant to open the pies. Charlie Batty, the same merchant, had approached several of his clients around Hull for potatoes, to no avail, which resulted in him desperately turning up at Fred's farm with a load of sacks, begging for potatoes. Fred smiled inwardly when Mr Batty disclosed, perhaps causing him to bite his tongue, that anything resembling a potato would sell in Hull. Realising the ball was in his court, Fred, too, pretended not to be interested, in spite of the realistic price he was being offered; but Mr Batty's final terms swayed him to go ahead.

The terms were a very stupendous price — no sorting, with just the sacks and potato scoops. Off they went to the pie, and Mr Batty helped to fill the sacks with potatoes and load up the lorry. His driver arrived later with another lorry. Mr Batty gave Fred a handsome cheque before setting off for Hull with his load. Subsequently, they came every day until Fred had not a potato on the farm.

Fred loved to sit with his pint in the local pub and recall his ups and downs and listen to those of his colleagues. One would have thought after a hard day amongst it all, they would have been eager to forget it for a while. Not so; and I must admit when I started going into the pub I never ever became bored with the same old banter, though they were all well enough aware that the faster the pints came in, the quicker the tales turned to fantasy. Beyond that, the conversations were on any mortal issue. Fred employed a horseman when his elder sister, Elsie, married and left home. This made life easier for his mother and Kathleen.

Mr and Mrs Victor Clayton and family farmed Sycamore Farm. They were in the milk trade in quite a big way by the standards of the time. They milked around twenty cows by hand, and each member of the family had their own little lot to regularly milk. Hand milking was regarded as a controversial issue, as cows vary in temperament, intelligence and genetic endowments, and they were, and still are, aware of one's ability, touch, approach, technique, patience and so on. Therefore, it was perhaps the done thing for a particular person to regularly milk certain cows, whilst other farmers may advocate and insist that his cows must be trained to be milked by anyone. But I do know it is not a fallacy to say a person whom a cow is used to can obtain more milk from it than could a stranger. Mrs Clayton was an excellent hand milker. She was such a gentle-natured lady, and always approached her cows in an assuring and respectable manner; but again, one might say that this soft, relaxed approach should be with the exception of, rather than the rule.

Their cows were a mixed herd; some were Roan Dairy Shorthorn, some Friesians, three very lovely Blue Roans, rather like the Belgian Blues of today, but with more blue evenly spread over their body, and the others would be crossbreeds. They grazed on the Asselby marsh in the summer and they looked so lovely with their new summer coats. Their colours contrasted with those of the meadow, which was a mass of buttercups, daisies, milkmaids, cowslips, red and white clovers, ragged robins and others.

These days such a pasture would be murdered without question. They managed it in the conventional way, with a heavy dressing of basic slag, plenty of farmyard manure chain harrowed in, and perhaps a balanced fertiliser in the spring. They would top it in late summer and create a thick second growth, and I never saw that marsh without a good swath of grazing on it.

Obviously, hand milking that many cows called for teamwork, and required at least two people at all times so they all mucked in when at all possible seven days per week, twice a day. It was very demanding and repetitive work, and they also bottled and retailed around the villages. Effie, the daughter, took charge of that section. Their dairy and cow houses were always spotlessly clean and maintained to the highest standard. Fortunately, they had the good sense and initiative not to let it become all bed and work. In the summer, they were finished by 6.30 p.m. with the cows back in the field, thus leaving them free to pursue their leisure activities. Dick and Bill, the sons, played cricket, snooker, league darts and were, in fact, into all sports. Sometimes my younger brother Gerald and I would go and watch them milking, if we were not elsewhere. Bill was a young eighteen-year-old and full of youthful mischief. He used to sing all the Bing Crosby hits as he milked the cows.

One evening as they milked, they were discussing who they thought might be called up for service when the war broke out, and Mrs Clayton said, 'If they call our Bill up for the army he will need to take a cart load of shredded wheat and a cow along with him.'

He used to drink the milk straight from the cooler, only a few minutes after being drawn from a cow. Such a practice would raise some eyebrows amongst the laboratory boffins these days, no doubt, although it has not done Bill any harm – he still plays a good round of golf today.

It was common for a horseman to show defiance should he have to help with the cows. However, Dick, the elder son made an exception here and did his share, but he was at heart a very competent and dedicated horseman, and highly skilled at breaking in the young ones. Some years he would have two in training, and bring them in from the field daily, put them in the pillar reins and get them going gradually, in amongst his other horse work. With endless patience and determination, he made sure they were thoroughly trained for all duties. The pride showed in his face when they were through it all; it was

## The Farms and the Farmers

like an army passing out parade to him. I was surprised to find, when I returned to Asselby to work, that they had packed up and moved, and gone all arable. I suspect it would be the seven days per week repetition that dairy cows called for that motivated them into that.

Their neighbours and relations were Mr and Mrs Stead on Riverside Lane Farm. Mrs Stead and Mrs Clayton were sisters. They too were in the milk trade. Mr Stead was yet another hard-working and self-made man. Most of his grass land was on the Howden marsh, therefore, he was up at 5 a.m. in the summer walking his cows to and from Howden. Looking back, this must have been a burden on the poor old cows and, indeed, went contrary to the theory of the experts, who have for some time advocated that walking cows twice a day over a similar distance would eliminate any profit therein. It goes without saying that Mr Stead did make money on that small farm, through his determination, and he was able to take on East End Farm when Mr Fletcher retired, and also later Sycamore Farm when the Claytons left. I would say this was quite remarkable progress in anyone's book.

Mr and Mrs Stead had quite a large family. The elder sons moved on to their own destiny. Frank became the foreman at Boothferry Farm, Armyn, for Mr R L Walker. This was a very prestigious farm, foremost in the local news. He later took on Field House Farm, following Mr Johnson. Herbert and Noel did their bit in the war. So young Charlie became his father's right-hand man. Now, Charlie lived up to his name's epithet, 'Cheerful Charlie'. He truly must have been the happiest young man in the village and he was quite a young gentleman, with the utmost respect for his elders, no matter who they were. I cannot ever remember seeing him without a smile on his face and his 'Norman Wisdom' style fits of laughter quickly spread to everyone around him.

One evening his father toured the village enquiring if anyone had seen Charlie anywhere, but no one had. He and his

*The Farms and the Farmers*

father had been putting in some very long days with an extra farm on board and Charlie had taken some horses to the grass field for the night. He decided to have a sit down on the grass on a lovely summer evening and fell asleep. It was here that his father found him, at eleven o'clock at night, sleeping soundly.

When the other lads returned from the war, it must have been some consolation to find their father had acquired two more farms, thus providing plenty of scope for them all. The estate built a new house to Riverside Lane Farm. Around 1940, it was the only house built in the decade I previously refer to. The old house stood about a metre from the lane and near the range of farm buildings.

Mr John Barker farmed Manor Farm, which was where he lived. He also farmed Ashgrove Farm where his brother, George, was the occupant and laboured for him. George had left Asselby and home as a young lad, no doubt in search of something new — a challenge, perhaps. He, in fact, had spent a long while down south and he never lost the southern accent. He also claimed his brother, John, who had taken over the family farm, had coaxed him back to Asselby under the pretext of setting up a dairy cow herd at Ashgrove, which he would have been in charge of, but it never materialised. George and his wife often had a good moan about it all. They also farmed Home Farm, which John's son, Harry, occupied.

The Barkers were private-minded people. Their farmyards and stack yards were always fenced and the gates kept shut, in some cases locked, and strictly 'no go' for us lads. They seemed to be farming more land than they could afford to, although some of their land was not known as the best in the village; and continuous corn growing, bearing in mind that there was no spraying, could lead to trouble on the poorer sands.

There was no Mrs Barker, so Mr John had a housekeeper, Miss Brackenberry, and a stockman living in with them. Taut Moore was his name, a very slow old man and, no doubt, he was employed for next to nothing. He milked the two house cows they kept, and in the summer he would sit under two very large chestnut trees opposite Manor Farm, on a Sunday afternoon, until time to milk. The trees looked beautiful in the spring when in full bloom with their candles, and they did, in fact, form the entrance to Mrs Taylor's orchard, which the Barkers used for grazing except in the fruit-gathering season. These magnificent trees, unfortunately, were felled as it was thought they posed a threat to Manor Farmhouse with their gigantic size. Old Taut was a flexible old chap and he would do household duties or anything for 'the Master', as he called Mr John.

Mr Joe Howden was a horseman for the Barkers. He had previously farmed Field House Farm, and came to grief in the treacherous 1920s. He kept on going to work until he was in his seventies. He was a pleasant old chap, and bore his regression with dignity.

Arthur Andrew and I came quite near to being guests of honour at Manor Farm, as we were invited to dine there on one particular day — due to unique circumstances, I must add. We were cycling to Barmby school on a very cold and frosty morning, wearing short trousers and not more than we needed or any other garments. On our handlebars swung our meagre midday meal of peanut butter, jam, or some similarly appetising sandwiches in a carrier bag. Then, the sight of a heap of thorn hedge cuttings briskly burning, with Mr John and George in their shirtsleeves eagerly adding more fuel, motivated us into dismounting, throwing down our cycles by the roadside, dinner and all and going to that lovely fire for a warm. After ten minutes or so and a chat to the men, they reckoned it was time we were off if we were to make school for nine o'clock. As we approached our cycles, we saw Ring, Mr John's collie dog, slyly walking away from where the cycles lay, licking his lips. We suspected from the bits of greaseproof paper littering the ground that he had undone and gulped

*The Farms and the Farmers*

down our sandwiches, and no one had noticed him in the act. So, we gingerly re-entered the field and informed Mr John of the matter. He and his brother showed no compassion and began saying, 'It looks like being a long day for you two,' and, 'Are you sure you had any dinner with you?'

Mr John called the dog to his side, took hold of his collar, and like speaking to a child, asked Ring what he had been up to. Then he gave him a pat and said, 'Well, lads, what is done is done, so you had better be at Manor Farm at 12.30 sharp, where you may dine with us.'

Due to these circumstances, we were late for school and had to explain to Mr Walker, the headmaster. He seemed none too pleased at the idea of us leaving the school at midday; perhaps he thought we were after half a day off, or it might have been his wife taking pity on us. However, Mr Walker informed us during the morning that his wife was preparing some lunch for us, so we never got to dinner at Manor Farm. Of course, Mr John asked the reason why, when we next saw him, but he smiled and said, 'I don't think Ring minded' – implying that he also ate the extra food Miss Brackenberry had prepared for us.

I clearly remember them hiring a contractor with this large machine called a Gyrotiller in the autumn of 1939. Now this machine was supposed to be reputed to be the answer to all arable husbandry problems, and half the village and many outsiders turned out to view it at work. The original models were steam powered, but they later became available with a crude diesel oil engine. The machine's main function was a horizontally rotating wheel, which penetrated the soil, with adjustable tines fitted as required to suit the various soils, and a selection of drive sprockets to coordinate with its speed in miles per hour.

It was late autumn, and had it been a clean field the results of the machine's ability would maybe have been to loosen the subsoil and thoroughly move and shake the topsoil, leaving it in large lumps enabling the frosts to penetrate, and so create a better spring tilth. This treatment would have been beneficial to the soil, but it was a warp land, absolutely matted up with wicks, or to be precise, couch grass. So their strategy was a disaster, as they were assuming the wicks would be gone by spring. With hindsight, I can envisage that it actually encouraged them to multiply. The infestation of couch grass on the warp or clay soils posed a major problem, and often called for a 'summer fallow' to correct it. Being a surface plant it could be dealt with more easily on the sand soils, and the most effective method of eradication, if not the cheapest, was to get them off by one means or another. Indeed, it was a common sight to see gangs of women hand-picking the wicks, or in very dry weather burning them in heaps on the fields, after the harrows had drawn them out of the soil. It required extra good land management and superb husbandry to keep one's land entirely free from couch grass, and there were plenty of farmers who could truthfully say their land was free from this troublesome weed.

The Barkers did not have all the wicks by any means, although George once said they only had one, which spread all over the farm. Harry kept Home Farm after his father retired, and acquired tractors and modern tackle that enabled him to farm to a better standard.

So, in those bygone days, Asselby comprised fifteen farms and twelve farmers who employed twenty-six regular workers. Additionally, there was always day work and piecework, available for both men and women, and strong boys and girls. Furthermore, the farmers formed an important clientele for the village shopkeepers and tradesmen. Therefore, they contributed to the whole of the rural society. Perhaps the rest of the villagers who had their own little enterprise could be termed as crofters or even entrepreneurs, as swings and roundabouts dominated their existence.

The Huttons lived at Westerby Gardens. Ernest and Morris

were the second generation of the family. They lived off about twenty acres of land, growing market garden crops, breeding and selling pigs, and a large square footage of glasshouses, which ensured they always had something to sell, thus keeping the cash trickling in.

Mr and Mrs George Whittaker lived at Thorn Lea Cottage, which they operated from, and survived on much less. Three orchards comprising five acres provided their living. Old George was a grafter. If one could have stood *Emmerdale*'s Seth Armstrong and George Whittaker side by side, they would have passed as twins. He was so laid-back, and utterly determined to let nothing worry or upset him. Yet, even though he worked for himself, George motivated himself conscientiously, and kept strictly to time. His opulent pocket watch and chain was his guardian, and he maintained it with pride and used to say, 'She runs like a Rolls-Royce.'

The two acres of land behind the house was planted with gooseberries and hazel, Honey Warden and Bell pear trees and others. There was not a weed in sight. In the winter, he would dig between the trees and bushes, and in the summer he would have loads of farmyard manure delivered, which he placed around the gooseberry bushes.

Whilst he was doing just this one day, he was approached by a representative trying to sell fertiliser, or 'tillage' as it was called in those days. Now George had it that tillage was for growing potatoes, turnips, corn, and the like, and promptly informed the representative of his beliefs.

'Oh,' replied the representative, 'we can make you a special!'

'Aye, I bet you can,' said George.

'Well, I can assure you,' replied the representative, 'a pocket full of our stuff will produce more gooseberries than a barrow load of yours!'

'How many more?' quipped George. 'The other pocket full?'

But joking apart, one can envisage how demanding his workload must have been. Imagine digging two acres with a spade, for instance; but it was not actually trench digging. He had a long-handled spade suitably set to just thinly turn over the surface soil, so from a near standing position he could cover quite a large patch per day. Their second orchard, behind the chapel, comprised loganberries, raspberries, and plum trees, and was kept equally tidy. In the back lane they had about one acre of open land, and some glasshouses in which they grew tomatoes and cucumbers.

George did not have a lot of interest in his open land, he always said the fruit paid better dividends, and he constantly declared that his greatest regret was that he did not have piped water on that section of his empire. Therefore, he had it all to wheel, he would add; but he methodically overcame the problem by making himself an unwritten rule, never to walk round there without a load of water, as he put it. So with a tank on his barrow, he did just that — whether he needed it there and then or not — and so kept a supply at hand.

Mrs Whittaker helped George at the fruit-picking season, but his strategy for marketing was so casual it was almost unbelievable. They would also employ a few women to pick the fruits, and many a time they would have a ton or so of gooseberries or whatever, picked into their own containers, with not a clue where they were destined for. He had a few contacts who he would write a letter to, or send a telegram, informing them of the situation and if he got the go-ahead then he would inform the LNER railway company to pick up his produce and despatch it from North or South Howden station. It seemed such a slow, chancy set-up, but it never worried George. Amid a cloud of 'Digger' flake tobacco smoke, he would happily discuss his returns with anyone who appeared to be interested. Sometimes he would risk a consignment of produce with the commission men who called weekly, but he always knew not to expect any miracles from them, as it was always on their terms. But it kept the ball rolling, George would say, and he often remarked, no matter how much anxiety his little enterprise created, it was worth a couple of pounds a week — less than the current weekly wage of an employer — not to have anyone to answer to.

Those very large Honey Warden pear trees were very prolific and would bear up to half a ton of fruit. He would harvest and store these, as they were as hard as iron when picked, but became excellent quality after being stored, and were sought after in the winter months.

Again, George was a keen advocate of being self-employed. 'Go for something of your own, my lad,' he would say, 'be independent, be your own boss.' Sound advice, no doubt, although his lifestyle as I describe it may sound to be mundane, boring, chancy or even insecure. If this was so, it certainly did not reflect in the faces of George and his lady, or in their personalities. They were happy with their lot. In their book, they had all they wanted. Many times through my life's ups and downs I have been envious of their contentment.

The music of the hammer on the blacksmith's anvil was beautiful to me even before I was old enough to ask what it was, being at the time mostly confined to the home yard. I can recall it being an almost daily sound, so full of mysticism to me. I have such vivid memories of the first few times I was allowed into the streets. There seemed to be so much going on, it was almost like a wonderland to me, and I believe one was fortunate if one had older brothers, as I had, to guide one through these early days, as mums perhaps did not always have the time.

Mr Tom Ellwood was the village blacksmith, and I refer to a highly skilled craftsman. He was, in fact, a master blacksmith and also a farrier. Being a placid-natured man added to his popularity, but at the same time, he tolerated no nonsense from anyone. When we consider the rigours of the long apprenticeship he'd had to endure to become what he was, he had the right to call the tune, and I think the farmers depended on him for the smooth running of their implements and machinery, rather than he on them for a living; but generally speaking there was a good atmosphere between them and they regarded each other as equal. Such was the hard old life of the blacksmith.

Tom's skill in shoeing horses was quite outstanding, and was very hard work, particularly with the young horses not familiar to it, and the very old ones, who tended to lean and transfer more of their weight on the poor old blacksmith when stood on three legs. When the ringing of the hammer stopped it was replaced by that low purring sound of the hand bellows, often operated by the whiskery old horsemen to warm up the forge, whilst the blacksmith held the shoe in the fire with his large tongs until it became red-hot and ready to be hammered into shape. The shoe was also tried on the horse's foot while still hot, to burn in an impression, and so ensure a perfect fitting shoe. Quite a cloud of smoke was created, which panicked the first-timers, and if the wind was in the right direction the unmistakeable aroma lingered around the village.

It might be assumed that a wet day gave the farmers the opportunity to get their horses re-shod without interrupting their land work schedule, but that did not always follow, and anyway Tom only shod by appointment, although he would perhaps shoe more horses on a wet day than otherwise. He did not appreciate too many onlookers, so if there was anyone waiting for their turn they knew to treat Tom with respect; but in the event of a particular horse playing up, they would all lend a hand and quickly master it. As a last resort, a twitch would be used on an awkward or timid horse. This was a piece of strong cord tied around the horse's top lip, and if it stood still it caused no pain, but otherwise pressure was applied by the man at the horse's head. In practice, this took the horse's attention away from the blacksmith's manoeuvres, transferring it to the pressure being applied at the twitch. The method would be looked upon as cruel by some, but the job had to be done, and positive steps had to be resorted to in the case of the latter. Anyway, it was beneficial if not essential that the horses were shod, otherwise they would have suffered worse pain through lameness and foot problems.

Another of Tom's major tasks was hooping the farm cart and wagon wheels. Now, this was craftsmanship of the highest degree, and whilst he worked mainly solo, this was a job he could not do on his own. These large iron hoops were formed from strips of iron straight from the foundry, and being the correct width, Tom would cut off a required length, and turn it into a circle on the tyre-bender. Then came the bit where he needed a striker to wield the big hammer whilst he held the hoop on the anvil, and so weld it into a complete circle. Young Fred, Tom's son, a full-time horseman, usually doubled for the striker, so his father had to choose evenings or Saturday afternoons, when Fred was usually available. Next came the very hot and sweat-inducing part of fitting the hoop onto the wheel. This was where Tom's brother, Fred, was involved, as he was the village joiner and wheelwright. He had made the wheel, and it was of paramount importance that both their measurements coordinated precisely. When they had accumulated a dozen or so wheels, hooping time was drawing near. They would not all be new wheels; some would be at the joiner's shop for repairs and re-hooping. So, again, when they were all available, which was nearly always on a Saturday afternoon, they would recruit one or two extra hands and set about some hooping.

The apparatus comprised a narrow, brick-built oven to heat up the hoops in, and a round iron platform slightly larger than the circumference of the wheel, which was a permanent fixture into the ground with a hole in the middle to accommodate the protruding wheel nave, and thus allow the wheel, when placed on it, to lay flat. This acted as a stop and indicated that the hoop, when hammered hard down over the wheel, was seated square and even the whole way round. The wheel was also fastened firmly down on the platform by a screw jack from the centre hole. A good supply of water, watering cans, and plenty of sledgehammers, tongs and crowbars were always at hand.

Scrap railway sleepers were used to fuel the oven, and with an intense heat built up, the hoop was placed inside the oven with the door closed and periodically checked and turned until the whole circumference was red-hot. With this achieved, two men with large tongs lifted it onto the placed wheel, which immediately created burning of the wood, and had to be slacked with a drop of water to avoid damaging the wheel's wooden components. The hoop was made fractionally smaller than the wheel, it having expanded while red-hot. It was hammered onto the wheel and doused with more water to contract it and so ensure a good, tight-fitting hoop. The men had to work fast amidst the steam and smoke to complete the job before the iron became cold. So one can imagine the conditions, combined with a hot summer's day, were quite demanding.

Tom coped with his umpteen other jobs with ease. In the winter he had the chisel and straight-toothed harrows to repair and make ready for the spring sowing time. Later a slow trickle of grass reapers and corn binders would appear near his shop, plus inter-row scrufflers, wick harrows, bow harrows etc. One very rarely saw a plough at the blacksmith's shop, as basically there is not much that can go wrong with a horse plough, except general wearing parts, which the ploughman, the farmer or some similar person could easily replace when required. These included points, shares, shin, mouldboards, landslides and wings, as the recognised wearing parts that are fastened to the plough's head and leg. If either of the latter became distorted or pulled out of line, a blacksmith would perhaps attempt to correct the fault, but a good ploughman would not be happy with it, and most bosses would be tempted to let the scrap man take it away.

*The Farms and the Farmers*

Despite Tom's heavy workload, he still found time for his hobby. Oh yes, he had one — one which many people would have given a miss! However, several half-days per week in the summer, his shop door was locked, and it had to be almost a matter of life or death before he could be persuaded to unlock it. In fact, it was a brave person who dared to ask him to do so. He had about five acres of land, in three plots, and this was where he could be found. On this land, he grew carrots, red beetroot, parsnips and sometimes a plot of peas.

He would hire a contractor to plough and work it ready for sowing, but the rest was down to him. He drilled his roots in 12" rows, which meant he could not even take a horse-drawn implement in for inter-row cleaning. So one can imagine what a task he had on his hands, and I must add there was no such thing as weedkiller or spray available at the time. If he caught a dry spell of weather, he would push a hoe between the rows, then creep up and down the rows, hand-pulling the other weeds, and his crops didn't look too bad. If it was a showery or wet spell at this critical cleaning time, he was in trouble, big trouble; but in either case, he loved it — it was almost therapeutic for him. As it slowly defeated him, he never lost hope or the will to battle on. This lonely figure would be just visible creeping amongst the rubbish, knee-high, eagerly tugging it up by the handful. One afternoon, as we walked home from school, us lads called out to him, asking what was supposed to be growing in the field. He gave us a wry smile, and replied that he did not know, as he had not found it yet!

When the pressure was on, Mrs Ellwood would help him out, but if she saw it becoming a waste of time she would plead with him to abandon it, and as they also kept the village pub she had not the time to waste. But Tom stayed devoted to his task, and she would follow him down the pub's path with a bottle of Forest Brown Ale, as he wheeled his cycle to the gate, setting off for another weeding session. Again, she would fuss over him and remind him not to be late home, or not to stay if it should rain, and many other imprecations a caring wife might utter. As he pedalled off, his last words would be something like, 'I'll be back in a fortnight,' or, 'Don't wait up!' – such was his sense of humour.

I think the first time I tasted Forest Brown Ale was one evening when we crept across his field headland, and the day before a bottle stood in the grass with only a mouthful or so taken from it. He had been so immersed in his work he had forgotten to drink it, let alone take the bottle home, so we each had a swig and ran off.

At the end of the year he usually had a few tons of produce to sell, which he was justly proud of; he had the right to be, for he put his whole heart into it, just like his blacksmith work. Having the village pub made for better relations between Tom's customers and himself. It was a meeting place or perhaps even an excuse for his current customers to get out for a drink, and at the same time discuss any forthcoming work they may require doing. Accounts would also be settled up openly amongst other drinkers with no qualms. The estate agent also took up office in the pub on rent days, and every six months the farmers, the cottagers, and so on congregated, some of them raking their pocket corners out, in order to 'pay up, and look pleasant', as they say.

The pub's main fare was John Smith's Bitter on tap, Whitbread's Forest Brown and Light Ales in one-pint screw-top bottles, Guinness, stout, a few bottles of spirits, and that would be about it. There was no bar, and no pumps, and most of the drinking was done in what was known as 'the kitchen', which contained a very large, long, beautifully scrubbed, white-topped table, with long settle seats, and two sturdy rocking chairs, which were strictly reserved for the landlord and his lady. It took a brave person to contemplate occupying one of those chairs. The spotless red-tiled floor contrasted well with the large black-leaded fire range, in which a fire roared even on the hottest day of summer.

*The Farms and the Farmers*

No, it was not to induce people to drink more, it was absolutely essential, because there was no electricity or gas. A constant supply of hot water and a hot oven was obtained only by a good fire in the grate. Particularly on a Sunday, the aroma of roast beef and Yorkshire pudding etc., floated around the place. Of course, this was not for sale, as no one had Sunday lunch in the pub in those days. One very rarely saw a woman in the pub, and never, ever, children. At Mother's request, we, as children, would knock on the back door on a Sunday lunchtime, and ask for a bottle of lemonade, and a bottle of beer to take out. Sometimes Mr Tom appeared, and perhaps having had a pint or two himself he would say, 'We don't sell that stuff here,' or, 'What do you think this is – a pub?'

We would stand there in silence, trying to work out what he was actually on about. Then Mrs Ellwood would appear, strike him with the oven cloth or something similar and say, 'Let them alone, you silly old fool! You are frightening them to death!' She would then take over the order, and this was when we observed what was going on in the place.

So in anybody's book, Mr and Mrs Ellwood did not have much spare time. Looking back, I often wonder what drove them on, but like many others they were happy with their lot. They soldiered on, seven days per week, 365 days per year. They never took a day off, never took a holiday, and never complained, and I dare say if they came back today and saw our carryings-on, given the choice they would take the one they left; they were such outstanding people.

As mentioned before, Mr Fred Ellwood, Tom's younger brother, was the village joiner, wheelwright and undertaker. A totally different man in many respects, but equally as skilled in his trade. He undertook a lot of work on the Knedlington Estate, as well as in his shop and bench work. Whilst he was a hard-working man, and dedicated to his trade, he was not a seven days per week man, and he would in fact be finished work, washed and changed by 5 p.m. during the week. If he had a bereavement on hand, and a coffin to make, at these times he worked extra hours to ensure it all coordinated. Perhaps through this part of Fred's business, Fred and his wife were regular church attendees. So every Sunday, sometimes twice a day, the BSA motorcycle and sidecar came out, and off they roared to Howden Church.

Mrs Fred was quite a domineering woman, but I suspect this was an occasion when Fred did not have to pretend to be deaf, because anything she might have had to say or moan about would be drowned by the old bike's noisy engine. If a customer visited the joiner's shop, and found him not available, as he had gone to an outside job, Mrs Ellwood would come out of the house, and leap to his defence, shrieking in her loud-pitched voice, 'Oh, he is so busy, he can't take on any more work for ages,' and all sorts of other remarks, almost implying she was his boss. But she was a nice enough woman at heart, and no doubt she was thinking of her husband's well-being. Anyway, her remarks bore no substance after all, for when she blurted it all out to him when he returned, he would listen in silence with an unruffled smile on his face, and they both knew that at the end of the day that all decisions and work rotas relating to Fred's work were down to him.

They had no children, so Mrs Ellwood referred to her hens as such. She kept 200 laying hens and was an expert egg producer. She spent endless time amongst those hens. The houses were scrubbed spotless, and she went to great lengths to ensure her hens were comfortable and content. She maintained these attributes were essential if one was to achieve profitable egg production. As they were free-range, she also kept a constant lookout for any roaming dogs, cats, rats, winged predators, or even humans, as she rigorously believed any sudden appearance from any of these did actually hinder the hens' laying performance.

The quality of those eggs deserved high praise as they were produced from nothing but the best feeds, containing no artificial additives whatsoever. Even in those days such eggs were a premium. Her outstanding skills in poultry husbandry were worthy of recognition, and I dare say they would have been beneficial to some of our not so thorough free-range producers of today, who perhaps might advocate such methods would prove unprofitable. It seems such a pity the profit motive has to govern all these days, and so rule out any other significant factor.

I suspect it was the last thing on those joiners' minds when they made those beautiful carts and wagons, that their redundancy was looming; but it was so, and it soon became inevitable that the ones they built in the Thirties would be the last, at least, in any great number or for commercial use on the farms. In fact, the rulley or dray, which was more widely used than the wagon in this part of Yorkshire, did survive for another decade or so. This was originally a four-wheeled horse-drawn flat vehicle, quite similar to the wagon but perhaps more flexible, as the sides and hay carting racks and shelvings could be fitted and removed as and when required. Some of them were modified for use with the tractors. Fred did make a few of these during the Forties and early Fifties, and as they were to be drawn by the tractors, he, like many other joiners, strayed a little from the original design; but as the tractors became larger, these too became unsuitable.

Consequently, the agricultural engineers stepped in and mass-produced a completely new model. Its basic dimensions and the materials it was constructed from differed vastly from the handmade ones. These were referred to as the 'farm trailer'. Today, even those have been replaced by the very popular two-wheeled tipping models, which are essential for the bulk handling methods of farm produce and more intensive farming methods.

Going back to when Fred was making a new cart or rulley, once again, Fred's brother, Tom the blacksmith, was involved. He made the iron parts, i.e. turntable rings, pole fittings, shaft fittings, stays, rope hooks, corner pins and axles etc. The axles, in particular, required a special skill, which most smiths held. They came from the foundry in two parts; therefore, the smiths had to weld them together at the centre to form a complete axle of the required length. The correct tempering of those axles was very critical in order to achieve the right consistency if they were to give a long and trouble-free service. The length of the axles also determined the width between the wheels of these vehicles, a crucial factor dominated by the county which one might be referring to, and the type of farming the cart or rulley was to serve, therefore, in this predominant arable area of the East Riding of Yorkshire.

The current width, or to put it into farming lore, 'the wheel centres', were strictly 56", the reason largely related to potato growing. Believe it or not, this is where Sir Walter Raleigh comes into this issue, for he brought the first humble spud to Ireland and soon they were cultivated throughout the country. I cannot pretend to know which preceded which, but I suspect after trials they discovered the potato gave the best results from being grown in 28" ridges, which became the current ridge width throughout the horse farming era. Therefore, the rulley and, in particular, the cart at 56" wheel centres fitted precisely down the centre of the ridges, when they were used up and down the same for planting and lifting purposes, and so eliminate any fouling of the ridge or damage to the potatoes.

So it goes without saying that whatever task Fred might be about to do, whether it be making a coffin, wagon door, window, gate or tumbril etc., it called for 100 per cent concentration and precise measurements and, with hindsight, I can now understand why he seemed a little abrupt, or even uptight when us lads visited his shop. He would always sell us a few pennyworth of nails or screws, and even give us the wood cut-offs which he had no use for, when we were making a rabbit hutch or a pigeon cage or whatever. He also allowed and appreciated it when we swept up the shavings and sawdust

*The Farms and the Farmers*

from his floor, which my father used for lighting his boiler fires; but, again, if our visits coincided with one of his busy days, his conversation would be very concise but civil, and if he walked to the shop door then we knew it was time we were on our way. Even so, he was not without a sense of humour: far from it! He was yet another hard-working, kind-hearted, good-natured, unbiased and genuine person, which one does not forget in a hurry.

*Village Life*

Life was, as it is today, no more complex than one made it. So many things have appeared which one could never have envisaged during my many decades of life. At the time of writing one does not always realise that the events and times to which I refer, and dwell on, are between fifty-five and sixty-five years ago. Some good and some not so good, and many the world would have remained a better place without.

There is a lot of sense, and indeed nonsense, spoken of the good old days. Even the opinions of those of us who were around sixty years ago and therefore saw those bygone days differ considerably. I regard the first decade of my life as the best, the most exciting, the least stressful, the most happy years; but having said this, there are so many aspects of one's life one must analyse before being able to truthfully state so. It goes without saying that the place I was born into, and took my first glimpse of life in, will remain forever uppermost in my memory. It projected such beauty; there prevailed unity, with an atmosphere amongst the people almost as if it were one large family. It seems so incredible that people — albeit not the same people I previously referred to — have brought about the social decline and destruction of the traditional image of these once farming villages, especially as the motive is, largely, money.

I can count the times on less than one hand I have been back into Asselby village since I finally left in 1958. Therefore, to the place and a large majority of the people, I am a stranger. Incidentally, I met, as I *returned* from one of these visits, what I would regard as strangers. Between Asselby and Knedlington, two large articulated lorries appeared on the horizon. No doubt they had minutes ago left the motorway, the M62, and they were using the unclassified road similarly. I gather these monstrosities can be up to 60 ft in overall length by 8 ft wide, and the road I refer to measures 17 ft 6 inches from grass verge to kerb edge. The average width of a saloon motor car, allowing for exterior mirrors, is 6 ft — thus leaving 3 ft 6 inches between passing vehicles.

I dare say that I am not the only car driver to be contemptuous when confronted with one of these awesome road hogs under such circumstances.

It was a great loss to people and rural surroundings when these solo owned and compact estates began to shed farms and properties. I suspect it was the very unjust death duties and other legislation which started the ball rolling, resulting in their rapid decline, and availability of middle-sized family farms to let. While some people saw these residential squires as dictators who ruled their villages and lands with an iron rod, I doubt if it really was so bad. Either way, at least it ensured a respectful and disciplined image, and it kept the right people in and the other kind out. Furthermore, the people had more respect for whatever laws they had to adhere to, simply because they were made by someone with the authority to do so.

This is more than can be said of our present-day bureaucratic council officials, blatantly dictating what one can and cannot do with and on one's own property, often quoting one-sided laws, which in realistic terms make little sense. Releasing properties from the estates also created, although perhaps not at the time, some very viable properties, i.e. farmsteads with no land attached to them.

These were offered for sale with all intents and purposes for the small pig and poultry keeper. However, it amazes me that the vendors or their agents did not anticipate that they held greater assets, for most of these divorced properties comprised in most cases a good farmhouse, barns, granaries,

Morris Parker ploughing for the next season's potato crop at a Lincolnshire Wold farm.

Harry Coatsworth at Ashgrove Farm, Asselby, returning from Goole's Agricultural Show.

A smarter looking Morris Parker, 1920.
(Note no harvesting operations taking place
– never on a Sunday in those days.)

Borough Blend, a Shire stallion, and groom prepare to tour the farms around Peterborough.

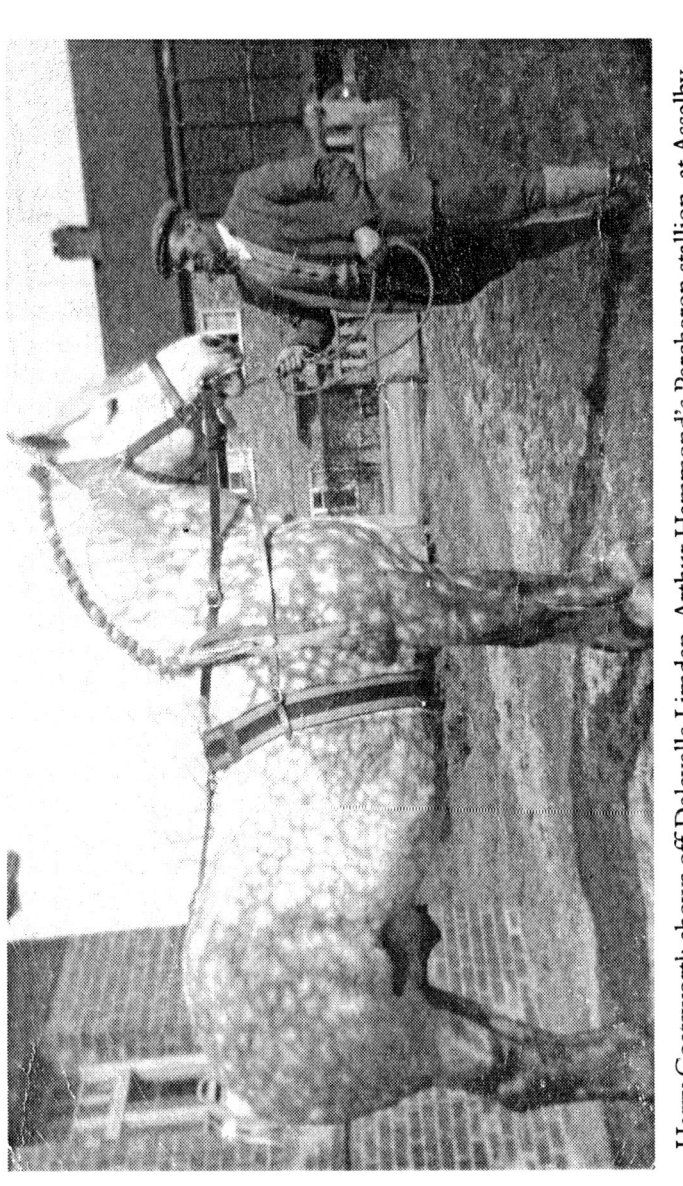

Harry Coatsworth shows off Delevelle Limden, Arthur Hammond's Percheron stallion, at Asselby.

The author leaves the blacksmith's shop. Blacksmith's Lane, Asselby. (Note the remains of the pinfold in the background.)

Frank Witty, Merv Walker, Jim Douglas and Dennis Ounsley prepare for some haymaking at Barmby on the Marsh.

Notts Crop Dryers Ltd, Misson, take delivery of fifty new Ford tractors – an annual event…

...with fifty new tractors and fifty one-year-old tractors to be driven away, one could say that the yard was quite busy!

The author, New Year's Eve, at Bridlington Spa Dance Hall, 1954.

A family wedding group of the Sykes and Clark families, at Linden House Farm, Hook.

Harry Drury turns his steam engine-hauled plough at Maugra Farm, Swinefleet Common...

...Harry keeps a watchful eye on his plough, which turned a 24" furrow up to 14" deep. Harry Drury became a partner with the Belton Brothers to form the Belton Brothers and Drury Company during the 1940s and took up the David Brown Tractor franchise.

The author, gun, dog, and fag, Elm Hurst Asselby.

The dogs decided to turn a blind eye.

Captain R E Sherwood parades with members of the Royal Family at Buckingham Palace for his and many of his colleagues' ceremonial recognition for their outstanding Second World War duties.

# BRITISH PRISONERS SET FREE.

Sons of Captain Sherwood, a Goole shipmaster, who were interned with their father on a prison ship at Hamburg. They have now reached home. They say they had enough black bread to last them a lifetime.

A much younger R E Sherwood (right), said to be one of the youngest prisoners of war ever after his ordeal at the outbreak of the 1914-1918 war.

First prize and champion at the 1910 Goole Christmas fatstock show, awarded to Robert Sykes, Ousefleet Pasture Farm. Sold for £43.10.0.

Grandfather's early twentieth-century potato lifting gang.

The author holds his first built corn stack. Father said it was a good job; the rest of it escaped the photographer. Sand Hill Farm, Swinefleet.

Eric Ellwood of Eel Hall Farm, Asselby, 1934.

*Tidy, an eight-year-old mare, and her twin foals owned by J. Ellwood and Sons, of Barmby by Marsh, near Howden, were among entries at Goole Horse Show on Saturday. (A Yorkshire Post picture.).*

Mr G E Ellwood and sons, who farmed at Asselby and Barmby on the Marsh, owned this very rare trio. Mr Ellwood had been involved with horses for some sixty years. The twin Shire foals were the first he had seen.

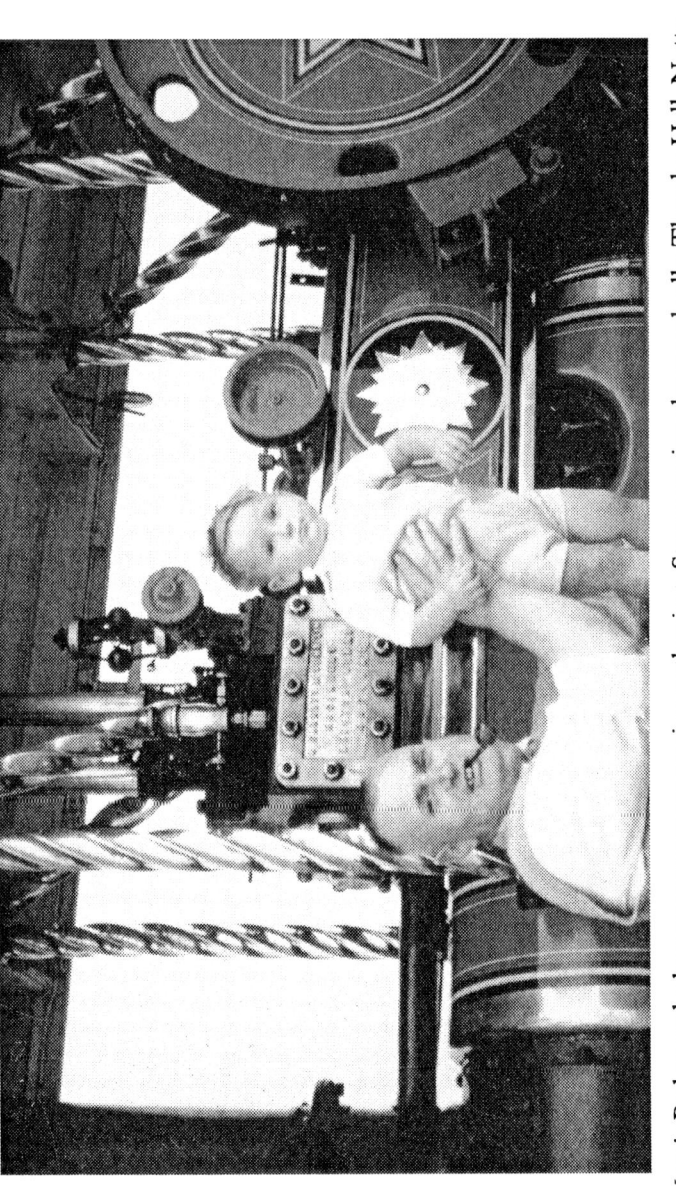

Morris Parker and a keen young steam engine enthusiast. Steam engine show and rally, Thorsby Hall, Notts.

Captain R E Sherwood tours his ship.

The author is still called upon to cut and lay many old and overgrown thorn hedges.

Note the curve in the hedgeline, known as a sneck in this part of Nottinghamshire. This dates back to the 1400s when Misson village was enclosed, the idea being to quell any animosity between landowners, tenants, OS mapping, etc.

The author and the co-owner of the hedge Mr Jack Bingham, a member of the oldest farming family of Misson, discuss the job in hand, 14 December 2005.

These layed hedges usually make loads of new growth resulting in a good thick fence in quick time.

The authors' grandson tries out the axe.

and other stock buildings, orchards, stockyards, fold yards etc., all of which are now being developed for dwellings, for which it appears planning consent comes as easy as buying a chocolate bar. True, we still have land held in large units, perhaps even larger than the former, but held by entirely different people. These large city money men and companies are an even greater threat to rural communities, the environment, and the well-being of the countryside. Local contenders cannot compete with them.

They are not interested in any such people; they are not even interested in farming per se. Money is their God, which they follow religiously. I believe my censorious view of the former evils would be mild against those of my older colleagues, and senior citizens, whom I have known and have now long gone. Should they ever come back to this topsy-turvy world – unfortunately we all know that day will never come, but just suppose they could – even an ordinary conversation would be for them somewhat confusing. Clothes, for example. The people to whom I refer were here before anoraks, cagoules, Barbour wax jackets, chinos, jeans, tights and drip-dry clothes; and briefs were, to them, something that barristers gave consideration to. If one mentioned 'Beatles', they would assume you were referring to insects, and not Volkswagen motor cars, or pop stars. Takeaway food, to them, would be food which was not eaten at one meal, but taken away and perhaps brought back for the next. Fast food was what they ate during Lent. 'Pot' was something they cooked in, and 'coke' was for heating.

Mention tea bags and they would assume one was referring to what the stuff was harvested into. Trainers, Ryders, Chelseas and Doc Martens would have them believing the conversation was pertaining to racing and not footwear! One could go on...

During the first decade or so of my life, when I lived at Asselby from 1933 to 1945, I cannot recall anyone taking a

*Village Life*

holiday, at least, not away from the village. If any of the farm men had holidays due, they would spend them working for someone else, or perhaps use the time catching up with work around their homes etc.; but it didn't follow that life was all bed and work — far from it.

One very rarely saw people working on a Saturday afternoon, and as for Sunday, it was strictly a rest day. The whole atmosphere and image of the place denoted this; everyone was dressed in their 'Sunday best', to quote a relative phrase. It was the day for attending chapel or church. On Sunday afternoon in particular the place seemed deserted; most people took a nap.

We were lectured and reminded of this on leaving the house after Sunday lunch, and generally speaking these unwritten rules our parents made had to be and were obeyed. And while it was not a hard and fast rule, five o'clock brought the end of the working weekday for people in most walks of life. It seems so incredible that after several decades of mechanisation and modern agricultural technology we have to tolerate the countryside, and in particular the roads, cluttered up with a vast army of large and noisy tractors and machines seven days per week, and often through the night, with little regard for people's feelings, rights and indeed safety. However, the aspects of the former and the latter differ as equal as chalk and cheese. In short, these present-day farm men could not tolerate such long working hours if it involved *manual work*, as opposed to the computerised machines and streamlined, stress-free tractors and so on which they are blessed with. This is why it was essential that the former worked to a more timely regime, simply because the onus was largely on people and animals, and not machines. This made, I believe, for a better quality of life than their fellow men.

Nights out was the 'in' thing. House parties, the picture houses or cinemas, concerts, whist and beetle drives, dances and social evenings formed the entertainment for people of all

*Village Life*

ages. When the outbreak of the Second World War occurred, people went in for the former with even greater enthusiasm, no doubt in a bid to take their minds off the sordid business. There might also have been a bit of 'Let the devil take tomorrow' in their minds, and the attitude 'Don't let Hitler get you down' prevailed. My mother, who was a trained singer, used to say the concerts were a bit nerve-wracking for her, and I suspect for others who got up there to do their bit in such meagre circumstances. These little shindigs were organised and advertised locally to attract visitors and artists from neighbouring villages, many of them happily travelled on their cycles in all weathers to participate. The concerts were held in the Methodist Chapel, which I never saw used as such. I think the only suitable attributes it offered was good seating. The rest was very basic; there was no stage lighting or indeed very little lighting of any sort due to the blackout. There was no accompaniment, except perhaps an old, out-of-tune piano, nowhere where the artists could wait in private for their turn to come up, no heating, no microphone and so on, so a very competent MC was at a premium. Despite the dismal atmosphere, everyone helped each other along, and there was always at least one of those characters in the audience capable of turning a raging thunderstorm into a bright and sunny day, once the ice was broken. All in all, on most occasions a good time was had by everyone. The whist drives were mainly adult events, while the beetle drives were more a communal affair. Anyone who could shake a dice and read what came up could be in at it; both were followed with a dance and social evening. These were held in the school room, a hip-roofed building, which now stands in near ruin. It made for better Mrs Walker's task, as she was the guardian of the place, which was next to their house, Elm Hurst.

So Mrs Walker and her colleagues organised these dos, which were much more civilised and homely. A large coal fire raged away in the black-leaded grate, the ladies mashed urns of

*Village Life*

tea and coffee, and despite the rationing prepared a beautiful buffet of home-made fare.

While the whist was in progress, us youngsters were not allowed access. After that, the doorman lowered his guard, took our few pence fee, and bingo – we were in there. We would help to clear the whist tables and arrange seating around the perimeter of the room, during which time the place quickly filled up. The MC would glare at his programme of events, which included all current games and dances for all ages, draw our attention to the same, then – *olé!* – two or three hours of wild riotous fun followed. There was always a portrait of the King and Queen in there, and I suppose the only thing which kept us conscious of the raging war was also a portrait of the very man who created it, Adolf Hitler. By the end of the evening it would be covered in hideous graffiti, with his name transcribed into more sinister traits, and no doubt it would be used to light the fire on the next occasion.

As for the nights out outside the village, Howden was the focal point. The climate provided plenty of scope for people from neighbouring villages, and most evenings it was like little London. There was, at the time, in, or quite near to Howden, the Women's Land Army Hostel, two air force bases in service, the flax factory, and lodging houses, where the gangs of Irish labourers stayed. They came here in droves, some of them year after year, and so they were well known by name, character, ability and so on to the locals. They took on contract work on the farms, at the fertiliser works, the drainage board etc.

Many of these young people serving in the former activities were regarded as possible rivals to the locals, consequently competition was at times fierce for both sexes, and in spite of the maxim 'All's fair in love and war', many a fracas broke out, which no doubt was more high-spirited bravado than intentional malice, as perhaps one-night stands due to the circumstances would be all these liaisons amounted to. The public houses were used largely for hard drinking and merrymaking, not eating. Sunday lunch in the pub with the family was a pipe dream; children and under eighteens were strictly forbidden in these places, accompanied or otherwise. With nine pubs in a relatively small town, it was a common sight to see these revellers moving from one to another, perhaps in search of the one with the best turn or crowd. Pub-crawling, I believe, was the common term for it. All the pubs had a piano and a pianist. Communal singing prevailed; there were also some very good solo singers around, who soon became well known and called upon to put their talents over.

Bernard Kelley was an outstanding tenor singer. Wherever he went he would be called for, and whilst he was performing the place would be in utter silence; I would say that in itself told us all we needed to know of his ability. The Drury sisters also regularly performed a beautiful duet; then there was old Jordie. I'm afraid it has to be said that anyone of the audience would have gladly bought her a half of bitter not to perform. However, no power on earth could prevent her from getting up there to perform her nightly ritual. The poor old lass would get hopelessly drunk every time she was out, and therefore totally unaware of the abuse which was being hurled at her, but at least she never caused trouble. She always seemed to sense when it was 'time she was away home', as she put it. So she would shuffle on, holding onto every wall, door, window or whatever, singing in such a dreadful tone. She was a lonely, pitiful sight, but of course it was the life she chose, and she seemed happy enough most of the time.

Sergeant Rex and his team maintained law and order in the place, and the fact that the police station was only a very short distance away from several of the pubs kept the landlords on their toes. Time was very strictly adhered to, which was no bad thing, as almost everyone had work to go to the next day. While I am not saying it was a whiter than white era, things which now seem to be the norm were viewed in a different light then. PC Nuttbrown and PC Drury concentrated on clearing the streets as quickly as possible after the pubs turned out. They tolerated no nonsense, and anyone who defied them was reminded they would be arrested for loitering if they continued to be abusive. Another fineable offence, for which people were given no second chance, was not having one's cycle lights up to standard. Two of my former friends, Albert Johnson and Maurice Hutton, appeared at Goole magistrates court charged with cycling under the station's subway on Boothferry Road, Goole, and consequently fined ten shillings each. Now, I do not know if the former offences are still on the books, but I do know that on many Friday evenings I visit a particular supermarket store, and here one finds youths and children charging around on unlit cycles on the footpath, swerving amongst people leaving and entering the store, and having no regard for the busy traffic-laden road adjacent to the store. I maintain that by relaxing their attention to these perhaps minor offences, the powers that be have sown the seeds that have established and caused crime to escalate into its present level.

Take the very frustrating spate of car thefts, for example, that we have to put up with these days. I personally am one of the thousands of people who have woken up to the shock of finding one's car, which one needed for work, missing. However, to get back to the mode of transport which fashioned most people's lives during the Second World War, 'the humble cycle': these too went through a phase when they were vulnerable to theft, or perhaps to the opportunist, and it was later discovered that many of these missing cycles went into the same watery grave, a farm pond opposite Breighton Airfield, where the Halifax bombers were stationed.

After the war, the pond was dredged for some reason or another, and this was when these forgotten rusty old objects appeared again, and no doubt solved the mysterious disappearance of them. It was no doubt sheer desperation which motivated these service lads or girls into snatching the cycles on impulse in a bid to get themselves out of their stranded dilemma. It could have been rear gunner Jones, bomb aimer Smith, wireless operator White, or co-pilot Brown, who stole Bill, Bob or Tom's cycle. Both parties were aware that this was hardly playing the right man, but I do know that civilians gallantly made exceptions in these cases. These very brave bomber crews risked their lives on our behalf; they were fighting for us all, they were fighting for their country. Some of them took off from Breighton nightly on bomber raids, determined to defeat the mighty German Luftwaffe, therefore no one paid much attention to a mere missing cycle; and anyway I feel I should add that the conduct of some civilians were not always impeccable, as the following saga proves.

One evening, the Bowman's Hotel at Howden was packed full of RAF personnel and others drinking, singing, flirting and doing all the things which pertain to a good old evening out. I am sure we are all aware of the common maxim, 'When drink is in, wit is out'! Perhaps this is what the landlord had in mind on the evening I refer to. But alas, at least one of his customers, a pilot officer in fact, proved to him that this did not always follow! After purchasing a round of drinks at the bar, he did not return to his friends at their table. Instead, he walked through the main door into the street with the tray of drinks. Unfortunately perhaps for him, the landlord, through a smoke-filled and very noisy atmosphere, caught a glimpse of him doing so.

He immediately realised this could mean trouble, so he quickly followed the man, caught up with him in the street, and promptly knocked the tray from his grip, sending the glasses of drink crashing to the pavement, and so proving his guilt beyond any reasonable doubt. Apparently the ladies of the group were drinking port wine and lemon.

Now, it was never proved whether the landlord ran out of port wine, or he deliberately served Vimto for the last half an hour, a popular soft drink at the time, which looked exactly like port wine. The man questioning this was convinced it was the latter; hence his intended trip to the police station. However the matter was soon forgotten, and it did the landlord's future trade no harm at all.

We were allowed to venture into Howden on a Saturday evening as we became older. I should be about ten years old when I first started doing so. The picture house was the main attraction. A sixpenny piece bought a front seat, but usually, due to the bus not coordinating, we entered halfway or so through the first house; that way, at least if it was a thriller, we could see the film over again from where it was when we entered, and if it was a cold night, we got more time in there. Our last penny would buy as many chips as a glutton could possibly eat.

Mrs Graybern's was our favourite chip shop, situated diagonally opposite the Britannia pub and the bus stop. The problem was getting in there, as there was a blackout. Mrs Graybern was constantly anxious that light was showing, therefore a large black sheet flanked the door and the window was totally blacked out. But that was not the worst of it. Once inside there was no order at all — the strong were served first and the weak whenever. They could cope with a steady trickle of customers, but a crowded shop for them was a nightmare. However, the end product was worth it all, believe me. It is no fallacy to say they were different, and if Mrs Graybern were there today, I would eagerly travel the thirty miles or so distance which I now live from Howden for a feed of them; they were definitely in the Harry Ramsden league.

One might argue, how can the common old fried fish and chips be different when they are still prepared from the same simple ingredients — potatoes, fish, fat and batter? I would simply reply that Mrs Graybern's were of a better quality. I used to really enjoy these visits to the small town of Howden. It gave us youngsters a chance to observe what was going on in the place, almost like dogs off our leads, wanting to be everywhere at once. The Bowman's, the Board Inn, the White Horse and the Wheatsheaf were the musical and more social pubs. The others in between were where one found the farming fraternity — in the Wellington and the Britannia mostly.

The Wellington was in fact the market pub, if you like, or the place where the farmers, the reps, merchants, and dealers would congregate to do business. Some of the lesser prominent farmers had nothing to sell, and nothing to buy anything with; for them, I suspect, it was an excuse to get into Howden for a few beers. It was a place that catered for one and all. Most of the business was done on the forecourt of the Wellington Hotel, actually in the street. Believe it or not, there were teetotallers amongst the farmers. Mr G Lapish and Mr Fletcher were in that league, and therefore never entered the pubs. As for those who were drinkers, I think it goes without saying when their business was done they lapsed into an innocent beer drinking session.

Mr G Ellwood, whom I later worked for and lived in service with, loved the afternoon out. He attended the market for both reasons. He often came home, not drunk, but merry, the better evil to bear for those who had not had any drink; he would have his wife, and me, in fits of laughter at the tea table, telling us what had been going off at the market.

The custom in many pubs at the time was to serve the beer in half pints, and the Britannia was a pub where this was widely practised. Mr George always said, 'You get drunk quicker on halves, the buggers come faster than pints!' He would then give way to an explosion of belly laughter.

Miss Edith Walker took over the Britannia public house from her father, and despite it being a very busy pub, she managed it single-handed for most of the time. She was a very tough lady and kept a disciplined house. Furthermore, no one argued for long with her when she picked up the stallion leader's stick, which she ruled her house with, in the event of any trouble. She had a philosophical way of dealing with trouble and its makers. As she once pointed out to me, 'It's pointless trying to bar a drunken man or woman – the time to act is when they come here again, quite often sober, and desperate for a hair of the dog... like now, for example,' she added, as two very rough-looking but sheepish Irish labourers entered the pub. They had given her a hard time the night before, and, proving their guilt, they tried to offer Miss Walker a feeble apology, and order two pints in the same breath.

She looked them straight in the eye and said fiercely, 'Now just you two blockheads listen to me! About last night, I want your assurance that nonsense like that will never occur again. If you cannot give me that, then please go away and do not ever come in here again... Now, do I make myself clear?' she concluded, in a voice as stern as that of a riled barrister.

They took the showdown honourably, and repeated several times, 'Very well, Miss Walker, we are sure you are right on all accounts.'

Their attitude satisfied her, and that was the end of the matter. I liked her style, and it proved to be not too high a price to pay, even to the likes of her victims, to not be deprived of such an excellent pint of John Smith's Bitter. After all, I would say most public house keepers accept that they have to be positive, but at the same time flexible in order to keep the customers rolling in. I suspect the reason why a lot of the landlords and landladies managed their pubs single-handed was the fact that their standards, for want of a better word, were simple, limited, but acceptable.

The whole system of the places was so basic, and so far removed from the complicated system of the modern pub. A glimpse of the layout of a present-day cellar, for example, would have the original cellarman frustrated, to say the least.

The Irish generally were a good, law-abiding, hard-working school of chaps, extra beneficial to the pubs, because they were mighty large capacity drinkers. Again, they came to Britain

*Village Life*

mainly to work on the farms. They took on work on a contract basis, and to a few English people who did the same they were regarded as rivals, mainly because they [the Irish] didn't pay income tax on their earnings. However, it was not a major problem; for one thing, there was plenty of work for them all, and I am afraid it has to be said, a lot of the farmers preferred the Irish. No one could deny they were very keen to get as much work as possible throughout the season. After all, money, and as much of it as they could possibly earn, was their ultimate aim. The same gangs, and in some cases the same families, would return every year, often bringing new ones along with them. Some of them had smallholdings back home in Ireland, which they left the wives and families in charge of, whilst they were here; 'crofters' I believe they were known as.

Prior to sugar beet being extensively grown in Great Britain, they would arrive just before harvest commenced. If they were a week or so early, or the weather was against starting the harvest, the large farmers would perhaps set them off cleaning the manure out of the fold yards onto the clover stubbles or fallows. One could not find much harder graft than that on a hot summer's day, but they really put their backs into it, and sweat constantly oozed out of them as they eagerly pulled and tugged at the aromatic manure, which was often four to five feet deep, and trodden solid by the cattle. Once the harvest was started, they worked between both jobs, which kept them fully employed until the last sheaf was pitched.

Of course, the campaign lasted longer than today's mechanised harvest, and if they had got off with a late start, it would be straight into potato picking on completion. Once the technique of sugar beet growing was perfected, the British Sugar Corporation induced more and more farmers into growing sugar beet. Consequently, almost all the arable farmers took up the growing of it. This was good news for the Irish, because it lengthened their work season; in short, they were required over here in early May, for the setting out, and hoeing and cleaning of the crop. They took it on at an agreed price per acre for a completed job, and whilst the agricultural wages board annually fixed a rate for all piecework, the Irish usually went for a few quid more, because they were aware of the fact that all contract work, especially outside, is subject to weather conditions. If a spell of wet weather coincided with this first operation in the sugar beet crop, they sometimes were in trouble, as it was essential, if not critical, that this first operation, namely 'striking out', was done as soon as possible once the plants reached the two-leaf stage. As Martin Cuff, the ganger, often remarked, 'These sugar beet plants are like the time and the tide — they wait for no man.'

Martin favoured working his men in one large gang, but at these times would split them up into small gangs, to keep the farmers happy, having taken on work at several farms. When the work amongst the beet and other root crops was completed, there was a lull on the work front until harvest time. Then, some of the married lads would pop back home for a few weeks, whilst the rest would toddle on with odd jobs here and there, and spend a few days at the races and so on.

Most of them came from Southern Ireland, and although they were not enemies they seemed to segregate themselves from those who were from the North. I also noticed too that they all had quite distinct features, i.e. the former were all tall, fair, fresh-faced chaps, unlike their colleagues, who were more stockily built, mainly dark-haired with sharper facial features, almost Roman-like, and a slightly different style of diction, with a lower baritone type of voice.

The sugar beet lifting — or 'pulling', as they referred to it — was very hard work, more so on the warp soil. This became their final job, and it often took them up to Christmas to complete it.

It is hard to believe, and it is a common sight these days, to see these fine and sophisticated sugar beet lifting machines piloted by their operators with speed, ease and perfection. By

*Village Life*

comparison those chaps did the job by handling each and every single beet. Their contract entailed pulling up the beet, knocking off the soil, laying them back on the ground, and finally topping them.

After a couple of months or so of this gruelling work, it began to take its toll, particularly on the older ones. However, the end product, which was a handsome wad, drove them on; and when they finally packed their cases and headed for the ferry, they were well worthy of the rest ahead of them.

Of course, the departure of the Irish didn't signal the end of the beet harvest by any means. The farmers' regular men picked up the beet (again by hand), loaded it into the trailers, and carted it off to a stockpile. It was advantageous for the Asselby and Barmby farmers to have the River Ouse bypass much of their land en route to Selby, which was where the local BSC factory was, so they would heap up a hundred tons or so on the river's foreshore, which was about the maximum barge load. The factory gave the farmer a permit, and a loading date — then yes! — it was all hands on deck. Furthermore, the task was governed by the tides, therefore the barge man normally favoured an early morning start.

While us farm chaps loaded the seven or eight barrows at the face of the pile, the barge men wheeled the beet on board along an 18' plank of wood, called the 'gantry'. It was dangerous if not impossible to load during a very high tide, because the position of the barge made the gantry too steep. A very low tide was awkward too, because it created the exact reverse effect. Alan Eastwood, the barge owner, would quip, 'I wouldn't mind if only the bloody barrows had brakes!' However, he and his mates were very keen, and understandably so, for they had other farmers' permit dates to meet or miss at their peril.

It took very hard conditions indeed to defeat them. Their worst hazard was wet soil or mud on the gantry, which was impossible to avoid. During wet weather, one man would often be detailed to deal with the problem. All in all, it was a challenge, and indeed fun, if not at times hilarious; and of course it was beneficial to us all to get a good consignment away at once.

The sugar beet grown nearer to the villages was conveyed to the factory by rail or road transport. Barmby railway station was fully operational, up to the late Fifties, where all farm produce was consigned in and out of the goods section. Mr Bob Lofthouse was the last residential stationmaster; he was a typical railway company jobsworth, but quite innocuous with it all. Ringrose, Hewsons, and Mashams, all haulage contractors, carted beet by road. Often they would arrive at the farms late at night, expecting to be loaded ready for an early morning flying start to the factory. So those of us who were not lucky enough to be elsewhere would light up two or three stable paraffin lamps, get out the forks and join the driver. It did not go down very well, after perhaps already having had a hard day. But needless to say, it had to be done, so the faster we worked the sooner we loaded the transport and got off to bed.

It was a welcome breakthrough for all sugar beet growing industry in the area, when Mr Maurice Shaw of Barmby on the Marsh bought the first sugar beet lifting machine to be operated in the area, in 1955, and set up his contracting business with it. It was a John Salmon single-row wonder machine in our book, although it was not very far advanced from the prototype. Any machine which would lift, top and load the beet into a trailer really stirred up some interest. Maurice bought a Ford Major tractor to attach the thing to and hey presto, he was away. Everyone seemed to want him at once. The contrivance took two men to operate it; one sat on the machine, steering the topping wheel by a long handle. This was independently aligned from the rest of the machine, although it still required precise driving, otherwise they were operating contrary to each other. Maurice, who incidentally rode on his machine for the whole of the first season, was constantly bawling at his tractor driver, 'Left a bit, right a bit!' and so on. However, as he often remarked, one could not be in a better position than that of sitting on top of the contraption, to observe what was going on with it. I was driving my tractor and trailer alongside them, catching the beet. I noticed he [Maurice] was often blaming the tractor driver for his own mistakes, quite unintentionally of course, for he could not resist the temptation to glance at the other moving parts of the machine, when his eyes should have been firmly fixed on the topping wheel. Maurice noted the machine's faults and quickly learned the A to Z of it.

One of the things which they realised they required and did not have was live PTO, a feature which was only half perfected on the original tractors. Therefore, this in itself put excessive stress and strain on the drive chains, sprockets, and webbings, etc., during a careless manoeuvre.

Fortunately perhaps for John Salmon, these findings were regarded as trivial by almost everyone who had been involved with the machine; and the farmers who had reluctantly allowed it into their sugar beet crops were very impressed with its maiden voyage, so to speak.

Sugar beet harvests progressed in leaps and bounds, and I appreciate it is so self-evident it need not be said. However, there is one thing I feel I must add. We are now well into 1994, and I see many part fields of sugar beet from the 1993 campaign still not lifted. I am afraid it has to be said, one never saw this kind of thing prior to the machine's appearance. The farmers I knew would carry the beet off the field one by one, rather than let the job beat them, regardless of the weather.

The smaller farmers who grew perhaps only four or five acres of sugar beet would attempt to manage the working amongst it themselves. If it posed a threat to them they would call us lads in to do the singling. Once we got a start, the temptation grew to miss school in order to earn extra cash; but alas, we dared not. The consequences would almost certainly have created an inquisition. Although the autumn half-term school break, which we used to call the potato-picking holiday, did become officially extendable for the final year lads, the government, the labour exchange and the school governors, all agreed to sanction this decision during the war in a bid to help the labour shortage. Any lads who wished to do so were given a kind of identity card to confirm to the farmer about to employ them that all was above board. This they in turn had to mark or sign daily, to prove that when the cards were handed to the headmaster by the lads at the end of each week, they had been where they claimed to have been.

The introduction of mono-germ sugar beet seed, pre-emergence sprays, band spraying, and selective weedkillers etc., cancelled out all the hand work amongst the root crops which formerly gave employment to all and sundry during the season. That was a consolation ever present in those days; there was always something going on in the village and always something to do, whether it be for love or money, it was there. However, the all too familiar but nevertheless true adage, 'All work and no play makes Jack a dull boy' did not prevail. But it is certainly true that strong young lads and girls had to contribute to the household budget, in kind if not in cash. So to quote Pam Eyres, the chores were there every day, and to repeat my own previous words, most households kept fowl, pigs, and cropped their gardens and orchards to full capacity, even more so when the outbreak of the Second World War brought about food rationing. Generally speaking, the chores which the father of the house did not get done before he left for his work were left to the rest of the family. Therefore, we often had jobs to do before setting out for school. This was no bad thing, and I am convinced the former circumstances instilled into our young minds a sense of responsibility and worthiness, and contributed largely to the fact that there was no such thing as gangs of bored teenagers hanging around the

*Village Life*

streets, inevitably bound for trouble. Furthermore, I am even more convinced that here lies the roots of today's muddled, mixed-up society, which seems totally without values; hence the spate of joyriding, petty stealing, and a general disregard for people's property, possessions or feelings, which we now have to tolerate. It is also not a fallacy when we constantly hear people stating the fact that unemployed adults also become bitter, bored, ruthless, frustrated, careless and in some cases deranged, ill and suicidal.

Mrs Palmer and her daughter, Anne, were not the same people after the outbreak of the Second World War. Running the general store and post office, they were normally cheerful, obliging, tolerant, flexible and so on. However, when the war brought about food rationing, and restrictions on just about all and sundry, they became uptight, frustrated and almost incompetent at running the same business. They hated having to turn people away, and perhaps tell white lies to those whom they knew had had their quota of a particular product, and the coupons flummoxed them; and I must admit us kids led them a dance with the sweet coupons.

Our mother flatly refused to have anything to do with them. She claimed, and I suppose rightly so, that she had enough problems with the more important and essential issues, so to those of us who were old enough to understand, she presented us with the sweet ration card, and wistfully added, 'Now just bear in mind, no coupons, no sweets; so don't come pestering me should the latter occur.'

I seem to recall the sweet ration cards being issued for five-monthly periods, each current month being indicated by one of the letters from A to E. If one adhered by the rules and regulations it was a simple and foolproof system. However, after boys and girls had their sweet coupons at their disposal, it became nothing short of a futures market. Wheeling and dealings took place in the playgrounds and other places.

*Village Life*

Coupons were exchanged for rabbits, pigeons, bantams, marbles and even money. We also discovered that the letter 'C' could, with care, be transformed into a 'D' and the 'D' into an 'E' etc.

But we had overlooked the fact that Mrs Palmer, or Anne, or any other shop, where one might tender the coupons, knew immediately, simply because one was supposed to hand over the card intact when making a purchase, and they took from the card however many units a certain purchase required. Obviously, the units detached from the card which were tendered were the ones that created arguments. Mrs Palmer would scold us and remind us we were breaking the law. She would threaten to call the local bobby and became beside herself. But she compromised in the end and, perhaps realising the units could only be spent once, would choose what order they were in.

Generally speaking, I would say the more essential goods had to be rationed, but sweet rationing seemed to me to be a farce. I don't recall any shortage, and it must have been equally frustrating for the shopkeepers having stocks that were not moving due to their being governed by coupons. Quite often, a group of boys and girls were gathered around the store window, longingly gazing through it at the jars of sweets on display... with a few pence to spend but no coupons.

The cigarette and tobacco situation posed a similar problem, though unlike the sweets they were not officially rationed but were in very short supply. Therefore I suppose one could say they were rationed to the retailers. One might also assume that in this case they were sold on a first come, first served basis; but this did not always follow. Mrs Palmer and perhaps many other shopkeepers always had a 'Sorry, no cigs' notice pinned on her shop door, even on the day she took delivery of her order. This was largely to ensure that her meagre supply went to local people, and not to strangers, or anyone who might be buying up for resale. There was nothing sinister in this; she was merely honouring her regular customers. We — the Clarks — were their neighbours, which was beneficial to us both; we were in fact some of their best customers, and very often, when our groceries were running low, we visited her shop for a packet of cornflakes or porridge, or a jar of jam at 7.30 a.m., when she would be lighting her fire and preparing for her nine o'clock opening time.

## *All at Land or Sea*

My father, Robert Arthur Gransville Clark, and my mother, Nellie, came to Asselby in 1931 and took over Asselby nurseries, a smallholding which was in a wild and run-down state. Dad had left the family's empire after a family feud, with little or next to no capital. Considering it was only two acres in all, he had ten good years there. We all mucked in, and were often left to it, when Dad also took up being a commission agent for a pea and potato merchant. Under glass, he grew tulips, chrysanthemums, and often flowers for the Christmas markets. These paid good dividends but called for constant surveillance of the temperatures of the glasshouses. If a cold spell of weather came, Dad would be out at 3 a.m. checking the solid fuel fires (which provided the hot water pipe heating system for the glasshouses) and ensuring all was in order; after all, a good sum of money was at stake.

We aimed at three crops per year by catching a quick crop of radish and lettuce before the cucumber and tomatoes went in. The latter sometimes had to be rushed on, in order to get the flowers in on time, as they had priority. In the open ground we propagated and grew flower and vegetable plants of all descriptions. There was always a large demand for brassica plants in those days.

I have a recording from one of my father's cash books which reads:

> 2,000 sprout plants to a Mr Arminson, a local farmer, paid with thanks £1.00, 5.6.1942.

There would be a steady trickle of such orders through the planting season. If one could estimate how many thousand plants could be grown from a pound of seed, even making a wild guess, there was profit within. So, between the sowing, weeding, harvesting, picking, watering etc., this labour-intensive little enterprise found us all a job, young and old. Incidentally, I had four brothers and three sisters.

The following are a few cash sales which I have picked out at random from another cash book:

    2.7.1939 – 47 lb tomatoes – £0 17s 7½d
    20.8.1939 – 10 doz. gladioli – £0 12s 6d
    9.9.1942 – 204 lb best tomatoes – £10 4s 0d
    3.6.1944 – 245 lb cabbage – £3 1s 6d
    2.10.1941 – 300 lb runner beans – £2 13s 9d

    3.6.1943 – Mr Blacker, Loftsholme Farm:
        1,000 broccoli plants
        3,000 sprouts
        300 celery
        500 cabbage – £2 11s 6d

    31.5.1945 – Mr Bailey, Knedlington:
        20,000 leek plants – £15 0s 0d

    10.11.1945 – Mr Heron, Asselby:
        800 wallflower plants – £1 13s 0d

    10.7.1944 – 90 doz. lettuce – £6 0s 0d
    16.11.1945 – 40 stn Bramley apples – £12 0s 0d
    3.4.1945 – 300 geraniums in pots – £10 0s 0d
    9.7.1944 – 116 doz. lettuce – £9 1s 6d

If we take the runner beans, for example, at two old pence and a silly fraction per pound, and compare the same product, which I saw priced at ninety new pence per pound today (10.8.1993), the latter in old money terms would be eighteen shillings per pound. It is perhaps for the better that we have no idea what they will cost in another fifty years' time!

The pea season was very hectic for Dad, and a far cry from the pea harvest of today. All the peas which were grown for the markets were pulled by hand. One would see as many as a hundred people, counting all and sundry, in the larger pea fields. The pickers were paid by the sack; the price per sack ranged between nine pence and two shillings at the time — subject to the fluctuation and the current market price of the peas, of course, which was very unsteady at times.

The whole operation called for close and constant cooperation between the growers, the merchants and the markets. If the market was slow, the price to the pickers would sometimes be cut during the day, which caused near riots at times. Even worse, the picking could be stopped altogether at a minute's notice; but I suppose the regular pickers were aware of these pitfalls. Two or three men were required to keep a large gang of pickers in order, and endeavour to keep them straight across the field in order not to miss any peas, as they tended to go only for the best if they could get away with it. Other men would be weighing and paying out as the pickers brought up their pulled sacks of peas.

Mr Blacker, Loftsholme Farm:
    1,000 broccoli plants
    3,000 sprouts
    300 celery
    500 cabbage

Mr Bailey, Knedlington:
    20,000 leek plants

Mr Heron, Asselby:
    800 wallflower plants
    90 doz. lettuce
    40 stn Bramley apples
    300 geraniums in pots
    116 doz. lettuce

Most of the peas went to their destination by rail, so they were transported to the nearest railway siding by horse and wagons. Perhaps the farmer would come out with the best deal, having let out the land for pea growing to the merchants, or the middlemen, as they were often known, for an agreed price per acre, with the money up front, so at least they knew what they were getting. The latter had the hassle and the risk to bear.

My father was involved with these types. Some of them were no better than playboys; there always seemed to be a lot of these characters around. They would perhaps be middle-sized farmers, or other similar business owners, and they would leave staff in charge of their businesses, whilst they almost daily attended produce and cattle markets, and inevitably spent hours in the pubs drinking and gambling, etc. Dad often went days without seeing the ones he operated for. He was up and off at 5 a.m., his first destination being the pea field or the farm where they had planned to pull peas that day. There he met whoever was in charge for the day, briefly discussing the plan for the day, and hand them a satchel full of cash to pay out to the pickers. Then he would board an express train and speed up to London, and other big markets, to oversee that end of the job.

It was a highly complex game, handling and organising the distribution of such a perishable product. Speed was the operative word, and a fresh sample of peas was the key to keeping the market men happy and in a buying mood. But even these aspects were subject to supply and demand.

Dad was a very energetic man. He loved to talk about his day's dramas. He would sit down to his evening meal, perhaps at eight o'clock or so, and tell us all about the day's events, which seemed never-ending. In fact, we would be falling asleep, and my mother would say, 'For God's sake, shut up and let them get to bed.'

Mother finally motivated him into packing it in after quite a few seasons. To be honest, there was not a lot in it for him, considering the long hours and responsibility he had, as he was expected to be fully accountable at the end of the day. Anyway, by the end of the mid-1940s, the static pea viners were booming, and they quickly superseded the green pea market. Indeed, due to a few factors, but mainly money and the lack of it, Dad decided to pack up altogether at Asselby and move on.

It seemed quite coincidental and simply by chance that we moved to a farm my father's grandfather had previously farmed. Dad and my eldest brother, Alan, took up posts as horsemen there. We occupied Sandhill Farm, which was a selection of what had become the Co-operative Wholesale Society Estate, formally Major Empson's estate. So while it was exciting and elating for us, it was nothing new to Dad, in a sense. He was moving into a house where he had spent his holidays on many occasions as a very young boy with his mother's family, the Sykeses, some thirty years ago.

His grandfather, Robert Sykes, had farmed Ousefleet Pastures Farm and Sandhill Farm, together with his sons George and Robert junior. So, to the amazement of the staff who were already there, he knew the place and its surroundings off by heart! Therefore none of them could tell him anything about the place, i.e. the field names and acreage, the road and the lane names, the landmarks and neighbouring farms and so on. He had been taught it all by his grandfather at a very young age.

He also remembered the days when Ousefleet Hall was in its full splendour, where a vast army of servants were employed to keep the large stately home and its surroundings in an immaculate state. It was Major Empson's seat at that time. These squires then ruled supreme in their manors and mansions, and were highly respected and appreciated. Sadly, it stood empty and neglected when we moved there in 1945. Jackdaws were nesting in the tall chimney stacks. Inside hung large cobwebs, and the beautiful covings and ornate staircases were showing signs of neglect.

I just could not comprehend how such a magnificent building deserved such a fate, but I dare say there were reasons; no doubt money would be the foremost, and needless to say it was demolished shortly afterwards. I suppose its nearest living relative is Ousefleet Hall Cottage, which still stands nearby.

It seemed like a dream to Dad to be back living in a house he knew so well, and despite the gap of thirty years or so, he would sit reminiscing and recalling events which had taken place in the house and in the farmyard as though they were yesterday. It gave him a tremendous boost to be back into farming again, albeit only as a workman; being a born farmer, it was nothing new to him, he had it all at his fingertips. Being born a farmer was an attribute which held a lot of recognition in those days. In fact those who were not could be ridiculed or regarded as mediocre by those who were. However, riches to rags is a well-known maxim in the farming fraternity, which is certainly significant in our family's case; for my grandfather, Arthur William Clark, and grandmother, née Hannah Sykes, were both descended from a long line of successful farming stock. My grandfather and grandmother farmed Lynden House Farm, and Manor Farm in the village of Hook, near Goole, where their three sons and one daughter were born and brought up.

My father, being the oldest child and perhaps the favourite in his father's eye, was educated at Goole Grammar School. He left at the age of fourteen to start his farming apprenticeship under the watchful eye of his father, and I gathered from Grandma on one of the occasions when her tongue was loose, that he was already theoretically brilliant on the subject. But I'm afraid it has to be said that she was to become his downfall. Her father (his grandfather) informed her of his potential and talent, which he had observed as the young lad of four years plodded after his grandfather around the farmyard, asking intelligent questions and never ever forgetting what he was told, on the occasions when they visited. 'Try him out — must see to it that he becomes a farmer,' he told his daughter, Hannah; and I suspect those words haunted her, for on her deathbed she repeated them many times. I, in fact, could not resist asking her why he was not, but of course she could not comprehend anymore, so all I got was a smile for an answer.

It seemed odd to me that she should never utter those words until the very end, especially when I already knew the answer; perhaps she had never been aware of that. However, as I say, Dad had yet to learn the practical and highly important side of becoming a farmer. So again under his father, and no doubt the head horseman, he learned to plough, row, sow, stack, thatch and above all, how to work, care for and generally maintain the workhorses, for they were their main source of power and therefore absolutely essential to the farm.

His luck seemed to be in, for after a very short time his father took on Decoy Farm at Rawcliffe Bridge. So Mam and Dad moved there, and he stayed behind to run the farms at Hook with a housekeeper to care for him. It goes without saying that he was absolutely delighted with the set-up and all was going well. Then, in 1918, his father was unfortunately confined to his bed with a serious illness, which sadly he never recovered from. This inevitably created huge problems.

They carried on as they were for a while, then of course it was decision time. Dad favoured continuing as they were, but his mother said no, she had no further interest in Decoy, and made it quite clear she was returning back to Hook. Now this was the last thing Dad wanted. However, she was his new boss, so to speak, and as they were two of a kind with regards to temperament, principles and stubbornness, and Grandfather was no longer there to act as mediator between them, trouble loomed.

He tried to get her to compromise by putting up the money for him to take on Decoy Farm on his own, leaving her a clear field at Hook, but again she was having none of it. Consequently, he damned her to his dying day. Mainly for such selfish arrogance, and largely because he knew a farm like Decoy had very great potential. Bearing in mind that the same farm sold for £775,000 quite recently, his judgement seems justified. So he had to eat humble pie and be downgraded, which he took very badly indeed. But that was not the worst of it by any means, for when Dad chose his future wife, further trouble erupted.

The outstanding fact was that Grandma Clark was a very strong-willed and stern lady. She'd had a very strict and rigorous upbringing relating to the Victorian age, and she believed in clearly defined values and customary beliefs. One of these was that the offspring of a farming family should preferably marry into farming stock. So it seems that Dad was adding fuel to a raging fire when he married my dear mother, Nellie, the daughter of Captain Robert Henry Hayes Sherwood, of Monrovia House, Mount Pleasant, Goole. A merchant marine captain, Captain Sherwood was a very highly respected gentleman in and around Goole. He was greeted by many people as he proudly walked home through the streets of the town in his full uniform after docking his ship.

He was a very tough and notable character, born into a seafaring family. His most extraordinary ritual was a daily cold water bath taken first thing every morning. From his first official trip at the age of fourteen, he spent his entire working life at sea, serving many years as an officer in Goole steam shipping vessels. After joining the company as second officer in 1906, he was promoted to first officer in the following year, becoming master two years later. He was a younger brother of Hull Trinity House, and had held a Humber pilot's licence since 1907. Up to his retirement, he served in many local vessels, including the *Blyth*, *Irwell*, *Berlin*, *Spen Altona* and *Liberty*; his last command was the *Dearne*.

No doubt his most hectic years were those of the 1914–1918 war, for he spent most of them as a prisoner of war in Germany. It was while serving as master of the *Equity* that he was taken prisoner at Hamburg, in August 1914. His two young sons, Robert and John, aged seven and eight years old, were also on board so they were also interned on the prison ship. Robert was in fact one of the youngest prisoners of war in the Great War.

So our Grandfather Sherwood had plenty of exciting tales to tell us wide-eyed youngsters. He was also shipwrecked on a sailing ship at a very young age, which he obviously survived. Grandmother Sherwood told us he often dreamed about it in his later years. She used to chuckle and say that he needed the whole of the bed to himself, and indeed the bedroom on those occasions. 'Oh, and the language!' she would add, with a touch of shame on her face.

They had a beautiful, exquisitely furnished house, of which they were justly proud. On his days off from the sea, Grandad kept the garden and the house's exterior immaculate. He was a keen bowls player, smoked the best cigars and he had a very opulent drinks cabinet. House parties were the in thing, as were card sessions of whist and bridge etc., which they frequently indulged in. It was exciting for us when Grandma asked us to stay for a few days, but we had to take our turn, being such a large family. I can clearly recall being utterly amazed on seeing, for the first time, a tap which ran hot water and a flush loo; such luxuries were still a few years away from our village of Asselby.

I think Grandpa's greatest disappointment was the fact that he had seven grandsons, none of which took to the sea. We all seemed so wrapped up in farming and rural village life. He would scoff and say, 'For heaven's sake, not another ruddy farmer!' when he asked each and every one of us as we approached school leaving age what our ambitions were.

Despite his ordeal with the Germans, young Robert followed his father's profession. It was almost inevitable that he would do so. He was to become Captain Robert Evan Sherwood, and have a fine and distinguished career at sea. From 1922 to 1926 he served his apprenticeship with the Edward Hands shipping company, and from 1927 to 1929 he was an officer in Court line vessels. He became a sub-lieutenant in the Royal Navy Reserve in 1929, and from 1931 to 1935 was mate and master in local ships. Through the

Second World War he served with the Royal Navy in command of trawlers on the Dover patrol, in corvettes and frigates on North Atlantic convoys, and in frigates on Indian Ocean convoys. The year 1945 saw him promoted to Commander, RNR, and for the last six months of the war, Commander of a Royal Navy air station. During 1943 he was awarded the DSO at Buckingham Palace in recognition of his work on escort duties in the Atlantic. He proudly appeared on, and participated in the television series, *World at War*. He looked so different to the Uncle Robert we had been used to when he visited us at Asselby on one of his short leaves during the war, with a full set – such a lovely black beard. However, Auntie Nora was none too thrilled with it and quipped, 'It's like sleeping with a tramp!' Also, I suspect that while she was justly proud of his outstanding achievements, she was pleased when in 1954 he finally agreed with her that the time was right to tie up his ship for the last time.

So Uncle Robert did just that, and shortly afterwards Captain Robert Sherwood joined the shipping division of British Railways as assistant marine superintendent. Then he was Marine Superintendent of the London Midlands Western and Scottish regions at Euston, and Chief Marine Superintendent of British Rail's shipping and international services division at Liverpool Street. He was the first British Rail marine officer to become Marine Superintendent. So it was not that Dad was marrying into a family of paupers, scoundrels or people without direction. But even so, Grandmother Clark remained adamant and gave Dad the 'Huh, fancy a farmer's son marrying a seafarer's daughter' treatment.

Consequently the families were never very close. They incidentally were regular attendants of Hook Church, every Sunday morning to be precise, where the atmosphere was quite frosty, to say the least. It was strictly 'Good morning' and the state of the weather terms; not that it bothered the

Sherwoods, though indeed Grandpa Sherwood took much the same view, and asked Nellie, who was to become my mother, 'Couldn't you find someone better than a common farmer's son to marry?'

Needless to say, Mam and Dad's life was destined for better times, had they hung in there; but when they left, Dad was the only loser, as Grandma Clark had other sons, but at least his pride was still intact. I maintain his mother broke a rule which must have been in her book, and that was the fact that traditionally the eldest son of a farming family came first in line for succession; this was generally and largely adhered to during that time. In short, she severed his primogeniture.

It seems to me to be one of those cases which highlights the saying, 'There's nowt so queer as folk.' Despite it all, Mother remained honourable to her ruthless mother-in-law, for she rarely mentioned anything about the saga. In fact it was Aunty Mary (Dad's sister) who filled me in with all the sordid details. She had been the young single sister who sat in the background taking it all in, perhaps wondering what unforeseen marital problems were in store for her. Nevertheless, she learned nothing from it, for believe it or not, she married a Londoner. This further antagonised her mother, who incidentally lived to see the occasion for, and no doubt utterly enjoyed using, the very popular words, 'I told you so,' when their marriage failed.

I feel I should add that Grandma Clark also had some fine attributes. It was unfortunate that her level of morals and standards were found overbearing by some people. She practised a wide range of traditional domestic skills, and would not tolerate slackness from anyone in that field. She kept household staff who quickly learned what was expected of them. One of the tasks she excelled in was referred to as 'siding a killed pig'. When a pig was killed, the ultimate aim was a minimum of waste. Therefore, when the carcass was dissected, the main parts, i.e. the hams, bacon sides and shoulders, were salt cured and dried, a relatively simple task. However, the rest of the meat at a glance looked a real problem, bearing in mind it could have been anything between a forty- and fifty-stone pig. It had to be processed into various pies, sausages etc., again with a minimum of delay. The better bits of meat were trimmed off the main limbs and made into pork pie and sausage meat; the large leaves of fat were rendered and the fat or lard stored; the head, feet and other related parts were made into brawn; the chines, chaps and spare ribs could, if the household comprised a good number of people, be used up as required; otherwise these too could be salted down and used later. So the only thing which was wasted, as the saying used to go, was the squeal.

She was also highly fashion-conscious and quick to criticise those around her who were not. She frequently took a train to London to buy clothes at the most opulent stores. Ladies were much more dependent upon public transport in her time, and woe betide any public servant who dared to ride roughshod over her, for she would retaliate with a few sharp and cutting expletives which would silence the most boisterous person. Her cuisine was of the highest quality, her etiquette reflected in her character, and even though I say so, she was a well-schooled lady. Incidentally, her mother was formerly a Miss Elizabeth Ducker of Graiselound.

Robert Sykes, my great-grandfather, was born at Crowle, in 1864. I expect he would be aware later on if not at the time, that the years of high farming, the golden age of Victorian agriculture, was about to bring prosperity to the British countryside. As it only flourished for two decades or so, perhaps he missed out there. However, it did not dampen his enthusiasm – far from it; for he started his farming career at Crowle on not many more acres of land than one could count on one hand.

I gather he was an excellent horseman and a very competent and meticulous farm man. These were attributes that stood any contender of the farming fraternity in a very good stead. His first relatively large move was when he took on, as a tenant, Sandhill Farm upon the estate of Major Empson, situated on the King's Causeway between Eastoft and Swinefleet.

Sandhill, a ring-fenced 150-acre farm, was a good proposition being good-bodied sand and silt warp soils made for better the working of it. Great-grandpa soon had 3½ pairs of good horses, a small team of men, a stack yard full of tidy corn stacks neatly thatched, and everything else that pertained to a good, hard-working, conscientious farmer.

In those days people noticed these things — not only ordinary people, who got pleasure and elation from viewing a tidy farmstead, and well-farmed and maintained fields; but also people who mattered, such as the Major. Yes, Major Empson, his landlord, had been discreetly observing Robert's potential, which he later was to discover very much to his advantage, I should add. So Robert, and his equally compatible wife Elizabeth, soldiered on, paying their way and living a reasonably affluent life. But like many others they were not amassing much money, to put it bluntly, and of course by this time they had four or five of their elder children to keep and educate.

But things were about to change. As he took his Sunday morning walk around the farm, accompanied by some of his children, this being the only day when he had the opportunity to do so, they met Major Empson. Now, at the time a neighbouring farm on the estate, namely Pasture Farm, had become vacant. Obviously the Major had glanced through his applicants, and he said to Robert, 'I see you have not put in for Pasture Farm then, Sykes?'

'No sir, I have not,' replied Robert.

'Is there any special reason?' exclaimed the Major.

'Yes, there is only one reason,' replied Robert, 'and that is the meagre state of my bank account.'

*All at Land or Sea*

'Well, I would like you to have it,' the Major went on. 'I very much like what I have seen of you up to now, so what would you say if I put up the money at the bank for you for the first year?'

'I would appreciate that immensely, sir,' said Robert.

'Done,' was the Major's instant reply. 'You can get on the land as soon as you wish,' he added, 'and I trust we will see you in my office on Monday morning sharp, where we shall sign up the agreement.'

They shook hands on the deal and went their separate ways.

The former was Great-grandpa Sykes's own description of that conversation between himself and the Major, passed down through the family. Does not such sagacity show what grand stuff the landlords and tenant farmers of those days were made of? The trust and mutual confidence they had in each other tells one everything about them, and at the same time it served to quell any misgivings either party might have. As he walked back home, Robert's mind went into overdrive. Had he done the right thing? Would Elizabeth scold him? Was it a good thing to be indebted to the Major? And so on... But his own intuition gave him confidence, especially when he found the whole of the family were delighted with the news. At the back of his mind he knew it would not be all plain sailing, but even so everything seemed right about the venture.

With 250 more acres of good silt warp land adjoining Sandhill merely by a few dykes, the older members of his family almost ready to join the crew, a larger house with a very attractive farmstead, and a very helpful landlord, one could say all he needed was luck, whatever that might mean. One of my old colleagues used to quip, 'Even bad luck is better than no luck at all.' So, according to him its meaning was immaterial, but I seem to recall that 'Good luck' was the norm and often the parting words amongst friends and neighbours of those days.

With the prospect of moving to such a fine place, the

Sykeses could hardly contain themselves, for unlike Sandhill, it was a more rambling house and of fashionable architecture, which meant it was divorced from the sights, smells, and goings-on of the farmyard, and a courtyard flanked the outhouses and service buildings. It had walled gardens facing gates at the south side, almost putting it in the class of the homes of gentry. Having said this though, it was many a farmer's philosophy that one does not get a living from a fine house; fair comment, I suppose, but for a farm of its size, these large houses were essential to accommodate not only the farmer's family, but also unmarried farm workers and household staff.

So when they moved to Pasture Farm, the head horseman moved into Sandhill House, and they worked both the farms as one unit, although they were both kept well stocked. Pasture Farm had to it a large acreage of grassland known as The Parks, which surrounded Ousefleet Hall, the estate's headquarters. Therefore Robert was able to carry more livestock. They would have up to fifty bullocks running in The Parks through the summer, plus the sheep; even though they were up to their knees in grass, they were fed cotton and linseed cake daily.

As a result they were drawing out fat bullocks and sheep for the market weekly. Robert always said this kept a trickle of money going into the bank to pay the wages and other incidentals through the summer. Once they were established at Pasture Farm, they moved by leaps and bounds, and they took on board yet another farm, namely Manor Farm, Adlingfleet, in a nearby parish. By then the older sons and daughters were working, and between them they established a flourishing family farming business. Consequently they could be termed as in the big league now, farming almost 1,000 acres. They became popular figures in the area, constantly winning first prizes at the Christmas fatstock show with bullocks and sheep.

Robert maintained and kept up his very particular image towards his arable farming section; only the very best work from the horseman would do. Despite farming being much the same as today as regards stability, they had some very good years.

When the eldest daughter, Hannah, who was to become my grandmother, married Arthur William Clark, it was a grand occasion. Robert was delighted when they chose the end of September for their big day, as it gave him the opportunity to proudly show off his home and its surroundings to the guests. Where else would such a proud man want the reception held, than Pasture Farm? So the number one priority was the corn, hay and clover stacks. They went to great pains to ensure they were correctly built and neatly thatched, and I have it from a good authority that they looked beautiful, almost model like. The wagons and carts were painted and parked precisely. As a good gesture, Major Empson's estate men painted up the farmhouse and buildings, and the ladies set about cleaning the large brick barn where they all were to congregate. These barns, with extra effort, could be transformed into acceptable banquet halls. The York stone floors were in some cases already polished with straw and hay etc. being constantly moved around on them. Once the temporary wainscotings were fixed, the rest of the task was straightforward.

The farmhouse ladies, too, would prepare the meal. The farmers' wives and their staff were excellent contrivers, and their cuisine was of the highest standards, on a par with the best hotels. It was of paramount importance to them that everything was right on the occasion. The onus was largely on the host and the hostess, therefore it was always better to be a jump ahead of the wooden spoon brigade − in the nicest possible way, of course. If only the camcorder had been around at the time to help the now silent old barns to reveal such splendour! The fashion, the style, the minute details of a nineteenth-century rural wedding reception taking place, in

*All at Land or Sea*

such a relatively unlikely place, could be accurately narrated by our parents and grandparents; anything further, unfortunately, was to be a dream.

So with Hannah gone from the Sykes household, the second daughter moved up in status. Lottie became next to Great-grandma Sykes in the household. Lottie later married into the equally determined, hard-working and self-made family, the Oades of Crowle, who are still flourishing today.

Farming had lapsed into a dreadful state at the turn of the twentieth century. Robert wistfully remarked that he wished that this was the worst of his troubles; but alas it was not, for his dear wife, Elizabeth, who incidentally was Miss Elizabeth Ducker of Graiselound before they were married, became ill, and sadly on 14 March 1911 she passed away.

Devastated as they were — Robert in particular — life had to go on. They supported each other, as families do through such setbacks. Perhaps Annie, the remaining and youngest daughter, was thrown in at the deep end, being left at the helm of the household so young. However, I dare say she got some consolation and support when she and their head horseman, Ernest Watson, who lived in service with them, became sweethearts. In the eyes of Robert, Ernest was a replica of himself, therefore fortunately they had his blessing.

The year 1912 was very wet and the harvest was virtually ruined through constant rain. The potatoes too were a near failure; they had planted 70 acres of them on the basis of a good previous year, only to sell about 100 tons in total. So I suspect as a result of such appalling circumstances they would be glad to see the back of 1912. Perhaps, had they known that more bad news was on the horizon, they would have lost the will to carry on.

However, it wasn't until 1918 that they were informed that the whole of the estate was for sale in one lot. They feared the worst, for it did not require a very astute man to figure what their chances were of continuing there. Obviously they were at

*All at Land or Sea*

the mercy of the new owners, and again bearing in mind how depressed farming in general had become, the odds were very high that it would almost certainly be an institution, and not an individual buyer, that they would have to renegotiate with.

They were spot on, for the CWS became the new owners of Empson Hall Estate. Robert's mind was made up immediately. He had no desire to be a tenant under the CWS; he said the individualism had gone, and he could not tolerate the prospect of having to deal with managers, directors, agents, etc. He suspected they would be subversive, which again he could not go along with, so it looked as though it was D-Day. Although Grandfather was a sprightly young man of seventy-two, he wished their departure had not been under such circumstances; but as there were six sons, they agreed it was the fairest way to wind up and each of them follow their own pursuits.

They had three separate sales, and Robert put as much enthusiasm into them as he had done when he kicked off with the farms. The best pair of horses at Pasture sale made 180 and 170 guineas each. Annie and Ernest married and started up farming at Newbrakes Farm, through wheels within wheels, I suspect — for the landlord was Mr Bennett, the owner of the Goole-based shipping company, and his wife was the sister of Great-grandfather Sykes, which perhaps proves the validity of my former statements. The third generation of the Watsons are still farming nearby to this day. A beautiful and opulent marble monument stands in Reedness churchyard bearing my great-grandfather and grandmother's names. As I stand beside it and reminisce all I have been told about them, I feel I knew them.

## Related Events

Recent statistics show that the less well off people are living shorter lives than their richer counterparts, with deaths from some causes being four times higher in the poorer neighbourhoods of Britain than they are in the richest.

I do not believe everyone dreams of being equally affluent, though it is clear that when a member of the less well off section makes a complaint or states a fact relating to this issue, Mr Major's Tories seem to surmise that is what they hanker after.

These findings are not going to go away or subside. Therefore if the government keeps on failing to recognise them, health, deprivation and the widening gap between the incomes of the rich and the poor will continue to escalate, and could in fact push the trend nearer to the bleak and perilous Twenties and Thirties Depression. Even so, I have always been an affable sort of chap, and I maintain when one gets past the threescore years stage in life, complacency rather than anxiety might determine how many more years one has left. Therefore one tends to pass on such issues.

But I do feel deeply concerned for the future of our young, growing-up generation, as I do believe all parents do, and largely for their sakes I very strongly believe that we should remain an independent nation and continue to make our own laws and set our own taxes, and to hell with the United States of Europe and its political and monetary union.

Why should the likes of Mr Delors rule Great Britain? They waffle on about world trade and us being left behind, isolated, and so on; all such propaganda needs nipping in the bud before it gets out of hand. The Kaiser of the German Empire set out to rule us in 1914, Adolf Hitler did the same in

1939, and almost succeeded in his attempt. Consequently a lot of very brave and good British people, and many others, went to their graves as a result of helping to prevent them doing so. God forbid the day when Mr Major, or any other politician, simply hands over our democracy to these power-crazed nobodies, otherwise they will have seriously failed Great Britain and its people.

I confess I have had, and still have, a lot to be thankful for in this life, and to be honest I cannot recall any hardships or unfortunate times from these so-called treacherous pre-war days; but they were there all right for some people, and I maintain the ones above governed one's lot when we refer to them. In other words, we country dwellers had much greater opportunities to tolerate or even combat poverty than our town dwelling counterparts. Although it seems incredible if not unbelievable that our standard of living is only slightly better than forty years ago, despite a 230 per cent rise in wages and salaries over that period, given that the annual budget is supposed to govern our revenue and expenditure, albeit subject to the fluctuations of inflation. Admittedly, these are highly complicated issues; but, even so, it seems no one has, as yet, mastered the rudiments of them.

I dare say the people involved would rather we were over-credulous than over-sceptical, should they wish us to speculate at all on the matter; perhaps they are not aware that most people's eyes see even more clearly when being hoodwinked. Who, I wonder, has benefited from all these changes thrust upon us? It certainly is not the people.

We were robbed immediately when the decimalisation system was introduced, for our shilling comprised 12 pence, not 10 pence. Value Added Tax, the dreaded Poll Tax, BST and others were introduced, merely to conform.

It is all a lot of bureaucratic nonsense from which we still await any appreciable results. Who knows, perhaps they were not meant to be beneficial to us?

*Related Events*

Believe it or not, there are many things which have changed during my lifetime which I applaud too. One which took place during the early 1940s and stands prominent in my mind was the acquisition of a school bus. Prior to that, we Asselby children walked or rode creaky old cycles to Barmby school. The new five-year-old starters rarely had cycles, and I have seen them crying with cold hands and legs on a winter's morning, as young boys never wore long trousers in those days; at times it was cruel, but as they say, one got used to it eventually.

Even deep snow did not excuse us from attending. It did, in fact, add excitement, plodding through and over the large drifts, although the teachers were at times quite concerned for us reaching home safely, if the snow continued to fall through the day. Of course there was great unity, not that they could pedal off and leave us to it. Under such conditions the elder members, family or not, always saw that the juniors were supervised.

Even though there were several bad winters during the late 1930s and early 1940s, with a lot of snow, the roads were kept clear, mainly by the farmers for their own sake, because they were up and down the roads daily, winter and summer. In the winter they were carting threshed corn, potatoes and fodder to Barmby Station, to be despatched wherever.

It was a lovely sight and a welcome one when the council showed up with the snowplough hauled by six large Shire horses. I believe they hired the horses from Shaws of Howden and other local farmers on these occasions. The horses' feet had to be sharpened, for obvious reasons, and if it was the wet kind of snow the horsemen had to periodically clean the hard packed snow from underneath the horses' feet, as it defeated the object of them being sharpened. The snowplough was a wedge-shaped version, so they could only effectively make one pass, to give the snow at one side of the road or the other a nudge further up the road. On their way back it was virtually impossible, as there was no tolerance to balance the thing, with only one side in work. The local roadmen would make passing places with their shovels, and it was beneficial to everyone to do their bit to keep the roads open, as there was also the horse-drawn grocery, coal and bread wagons to consider.

Of course the mothers were concerned, when their — in some cases — young and innocent five-year-olds were faced with the 1½-mile walk to and from Barmby school, often in near darkness in the winter, particularly the ones with no brothers or sisters. But then again, there were plenty of others to tag along with. Even so, one very rarely saw mothers accompany their children, which beggars belief, and tells us much about the times we live in. Nowadays, for the sake of the children's safety, the same journey to and fro that a lot of children have to endure now calls for strict vigilance from parents and teachers. I notice particularly in the towns where relatively short distances are still walked by the pupils.

Mr G F Walker was the residential headmaster of Barmby school in those days, with Miss Joy next to him, and Miss Chantry taught the infants. Mr Walker was a Leeds man, and while he held the post for two decades or so, he was never very popular. In his ways, his beliefs and his style, he remained a 'town man'.

He very rarely entered the village, and took no interest or part in it, nor in its out-of-school activities. He often ridiculed the people of Barmby and criticised the way they spoke, the way they lived and so on. I suspect when a total stranger moves in amongst a close-knit community of people as a figurehead, one has to run with the fox and hunt with the hounds if one hankers after the best of both worlds.

However, he was a good and conscientious headmaster, who did his job as amicably as he felt obliged to, but in an ominous manner. Mr R Everatt, Mr C Leighton and Mr G Ellwood, three village farmers, were also school governors, and also trustees for the Mr Green Charity, which had left for the school. Therefore Mr Walker had to liaise with them, albeit reluctantly.

Mr Walker would often give us a running commentary of the farcical goings-on he had had to endure at the previous night's meeting; according to him it was he who spoke all the sense, and they who spoke all the nonsense; but with hindsight I can now envisage that those three jolly farmers deliberately set out to antagonise Mr Walker. However, I dare say that a line would have to be drawn somewhere with this playful banter and badinage. Even so, there was no way Mr Walker could put those men down, with all his fancy words, for they didn't give a hoot for anyone, let alone the school's headmaster.

Mr Everatt had a habit of borrowing a few of Mr Walker's pupils during school hours, sometimes without his consent, despite his loud protests. When Mr Everatt had a large herd of sheep or cattle to move from field to field or wherever, obviously as many hands as possible were needed for the job, so Mr Everatt would recruit a few strong lads as they walked home for lunch, and he was not above coming into the playground whilst Mr Walker was having lunch.

However, Mr Everatt was always honourable to his tormented colleague, for he always brought the lads back to school in his old car, often an hour late. Mr Walker would meet, him ranting and raving. Having heard it all before, Mr Everatt would stand there with a wide grin on his face waiting for the last word. He would then respond by saying, 'Well Mr Walker, it will be a shorter afternoon for them now!' or, 'Don't tell me you actually missed them!'

On one occasion he quipped, 'Oh, by the way, I forgot to pay them.' Taking from his pocket a penny piece, he handed it to Mr Walker, and asked if he would share it between the lads. Of course the lads in question were in for a rocket, as he was unaware that there had been a 'quid pro quo' between the lads and Mr Everatt. Mr Walker would in fact give us all an assembled lecture on the pitfalls of our even contemplating working for these penny-pinching peasants on leaving school!

I suppose Barmby Council School was in a way unique, for in a sense, one got paid for attending it. The 'Green Charity' which I previously referred to was left by the aforesaid to the school. The interest from it was to be shared between the pupils each Christmas. As it was based on attendance, it also created some incentive. The maximum sum was ten shillings, which a few pupils received. Bearing in mind that this called for 100 per cent, or one year's full attendance, barring official holidays, it was a remarkable achievement, and ten shillings in those days was a very appreciable Christmas box, albeit its equivalent today (fifty pence) would not, as they say, be worth getting out of bed for.

I remember clearly the last time I had to march up to the table, bearing each pupil's lot, neatly stacked in rows to receive my 'seven bob', as the clerk termed it, when calling out our names to do so. It was 1944; I can recall us all being assembled there, waiting for Mr Ellwood to arrive before the ceremony could commence. He and Walker shook hands when he arrived, and one would have thought that knowing what he would be up against, Mr Walker would have left it at that. However, he went on to say to Mr Ellwood, 'You are looking well, but then you are living off the fat of the land, so it is what one would expect, isn't it!'

'I don't know about that,' replied Mr Ellwood, 'I am not as fortunate as you, sat in here all day with nothing to do.'

'Ha,' replied Mr Walker, 'I wish you had my job to contend with!'

'Yes, so do I,' said Mr Ellwood, followed with a chortle. 'If only you were worthy of a better one, then we should both be better off!'

Fortunately for Mr Walker, who was struggling to compose himself, one of the other presiding gentlemen proposed that they should get on with their intended duties, which they promptly did.

*Related Events*

I can recall another time when Mr Walker got a bee in his bonnet, believe it or not, over the humble spud. He asked one of the lads if he could acquire for him a sack of potatoes from one of the village farmers. The lad came to school the next morning and told Mr Walker that he had asked every farmer in the village, whom he knew had potatoes, and they had all given him the same answer, which was, 'We 'a' noan up!' This meant, when translated from farming terms into plain English, 'We have not got any up.' Now, there was a perfectly logical explanation for this, but Mr Walker chose to believe they were deliberately refusing to sell him potatoes.

So once again we got a lecture on the peevish and selfish ways of the village people, and so on. Today a layman, as Mr Walker was, would not be up against such a complicated issue, for the storage methods for potatoes differ vastly; hence the 'Potatoes for sale' signs we see at farm gates up and down the country almost throughout the year. Formerly, all potatoes were stored in pies or clamps out in the fields covered with up to three feet of straw and soil to protect them from the winter weather. Therefore if the winter weather was severe, or the trade slow, the farmers were very reluctant to open up the pies, especially for a small lot. Despite Mr Walker's assumptions, the farmer was actually telling the truth when answering the lad's enquiry with, 'We 'a' noan up!'

Miss Joy and Miss Chantry were both country born and bred people, so I suspect towards Mr Walker's antagonism they held an open heart. Miss Joy was in fact the daughter of Mr Will Joy, Mr Walker's predecessor. She was a loud and devout woman, absolutely in full command of her job. She ruled her pupils with a cane, and a right hand as fast as that of Mike Tyson. It was nothing short of brutality. Some of her personal habits would never have been accepted in this day and age.

She would dine alone in her classroom with the door locked, and consume a daily diet of Marmite sandwiches and Woodbine cigarettes in front of a roaring coal fire. She wore thick tweed clothes in the winter and never discarded any of them, even under these circumstances, and I should add that she lived in a house with no bathroom. Need I say more? One can imagine the aroma in there come one o'clock, when we had to march back in there and resume lessons.

Mr Walker's face revealed all of his inner thoughts on the odd occasion when he had to enter her room. He in fact rarely did so. He would call her to the open door, and stand upwind of her, keeping his mission as concise as possible. I must add to that the old dear had caring traits, one of which was that she always ensured us pupils had a packed lunch, and got a hot beverage of some kind. She had this large iron kettle which held at least two gallons. Once we had filled it with water she took sole charge of it; she absolutely forbade any pupil of any age to handle it, it was her strict ultimatum. So before she settled down to her own lunch she would lift the boiling giant onto its stand, and forward we went for our fill. One always knew that if she stopped pouring someone was breaking the other iron rule, which was always keep a safe distance between each other.

Another of her strange habits greatly fuelled our imaginations. Now, whether it was to protect us from vanity, or her embarrassment, we never knew, because no one dared to refer to the issue. Perhaps it was a little of both. However, she was an avid reader of the magazines, *Picture Post* and the *Illustrated*, which contained a mishmash of everyday rural and urban issues, just as do similar ones today.

She in fact often quoted articles from them and suggested that our reading of them would be beneficial to our learning in many ways. But alas, before we were allowed access to them she carefully read and observed what she thought was suitable for our eyes and heads, and what was not. Hence by the time they were placed on the desk for our benefit, several of the pages would be glued firmly together, and no one knew better than we did that it was an utter impossibility to part them. Whatever it was on those pages, which Miss Joy chose to

*Related Events*

protect us from, again we never knew; it might have been an early version of our present-day Page Three *Sun* girls, or perhaps the original Marge Proops column, or even merely advertising women's underwear – I do not know.

Even so, I am sure Miss Joy had only our best interests at heart. She strongly advocated that children must be taught and raised with healthy bodies and minds. A good standard of morality was at the top of her agenda. I suspect today's experts on the issue would condemn and regard Miss Joy's actions as doctrinaire, but I would say at the end of the day she taught us more about immorality than their permissive liberalism could ever do.

Miss Chantry was a totally different character from her colleague, Miss Joy. She was a quietly spoken lady with endless patience, entirely suitable to teach the infants. I clearly remember starting school after the Easter break of 1938, and finding it difficult to stay in one place for such a long time. However, with her patience and charm Miss Chantry was well enough aware of these symptoms, and quite capable of expertly guiding one through them, and so giving the new starters confidence to settle in and conform.

When occasionally the old dear did become uptight it would upset her greatly. Her face would become bright red as she ranted on, followed by a deathly white contrast. She then would have to make a hurried dash to the toilet. I say 'old' because she was nearing retirement age when I started school, and even as young as I was, I could detect that she was under pressure at times. But worse problems were looming, and not just for Miss Chantry, as unfortunately the whole school was about to be thrown into turmoil. By now the Second World War was raging on, and many cities were evacuating people, mainly children, in the hope of getting them away from the wrath of Hitler's merciless bombing raids. So overnight, the population of Barmby school's pupils increased by approximately thirty.

*Related Events*

I will say pupils, because some of these new faces could hardly be referred to as children, being in their early teens. In any event this was not particularly good news for a school already carrying a maximum load. However, there was no evading the problem. It was well and truly there, and it had to be dealt with. Because no one seemed ready for it, it was absolute chaos on the Monday morning when they all turned up at Barmby school.

It almost looked as though another war was about to start as we all walked to school. It was very much a case of us and them; the atmosphere was very menacing. There were some big rough lads amongst them, and they were in a pretty mixed-up mood, as no doubt we would have been had we been subjected to similar treatment. They were already missing their parents, families and their own environment etc., and they were none too sure what kind of world they had been plunged into. Although they were not so far away from home, being from Hull, the wrench for them was hard to bear. There were also some rough and fearless lads amongst the Asselby crew too, but at least common sense prevailed on that first morning. Just for the hell of it, I suppose, or perhaps it was the first time they had been so near to them growing, these Hull lads had armed themselves with a stout stick each, which they had cut from the hedgerows alongside the road. This was hardly a show of unity.

Mr Walker always had his hair brushed flat down with some kind of gel, and there was always a tuft that stubbornly stood up. Not long after realising what he was about to be up against, with his new impudent guests, it was almost all stood up. They hung onto their sticks most of that first day. Mr Walker's number one problem was room – or the lack of it: where was he going to put them? All these questions Mr Walker had to contemplate.

But as in all aspects, when a war breaks out one has to make the best of any predicament. The first week was hectic for all the teachers. Mr Walker was not used to being challenged or answered back in any other than a routine way, and he was getting all this and more from them. He became very uptight and frustrated, especially as it was happening when we regulars were present.

However, after a few weeks of muddling along, things took a turn for the better. Miss Chantry decided to retire gracefully – she could stand no more of it, and Mr Walker acquired three more teachers, who were in fact themselves evacuees. Miss Albion, Miss Robinson and Mr Trip all found temporary homes in Asselby and lightened Mr Walker's burden considerably.

They divided the classrooms with ropes and curtains, and between them very quickly had law and order restored in the school. Out of school, the atmosphere remained frosty between us and our rivals. One morning whilst it was still gang tactics, the evacuees formed a human barrier across the road as some of the older Asselby lads cycled up behind them. I was riding on the crossbar of my elder brother Alan's cycle. But alas, they showed no fear and unanimously decided to charge through come what may. This resulted in an ugly scene, the cyclists being outnumbered.

Fortunately for us, some farm men were turning their teams of plough horses near by. One big six-foot fellow observed what was going on, and others intervened. Alan Everatt, the big fellow, collared a lad in each hand and threatened to bang their heads together. Between them they literally scared the living daylights out of the tormentors, quickly bringing the fracas to an end. Gradually they realised that we were not averse to them, and friendliness prevailed. It also seemed to me that despite the circumstances, they had come to better pastures; in other words, it was home from home for them, especially the ones who were placed with people with no children of their own. George and Pat Wilkinson were two such lads; they were placed with Mr and Mrs Lapish of Croft Farm.

*Related Events*

Now, I cannot pretend to know whether it was a burden or a delight for these people in their mid-fifties to be suddenly presented with two boisterous young boys. But I suspect it would be the latter, as everyone felt it was their duty to do something towards the war. After all, those unfortunate young children were blameless. Mrs Lapish did actually fuss over them, and she showed great concern about where they were out of school hours. She once said to my mother, cheerfully of course, 'Now I know how you are fixed with eight of them to ponder over.'

Tom, Susan and Jimmy Wilson were also similarly placed with Mr and Mrs Barker of Ashgrove Farm. Their only son, Jack, was deaf and dumb, therefore these temporary additions to their family were great company for them.

The Barkers did their utmost to make them feel at home. Mrs Barker had in fact gone through similar circumstances herself during the 1914 war, therefore she knew what it was all about. There were only the Cooks, the Riaheys, the Wards, the Bells and the Browns, who I can recall. The Browns managed to get the tenancy of Rose Cottage, therefore the whole of their family came to Asselby, and were indeed reluctant to go back to Hull when it was all over; but they did eventually do so. Mr Trip also acquired Wayside Cottage, enabling him to be reunited with his wife. Some of their children were in fact born at Asselby.

The Second World War brought scenes and events to Asselby which otherwise would never have happened. Sunday, 3 September 1939, remains etched in my mind as the day I was enlightened about something mystical, that simple word 'war' with its mammoth meaning, which had scarcely been previously mentioned that I could recall.

As we sat down to tea that Sunday evening, my father took up the subject which he had learned from the radio bulletin earlier in the day, which astounded the whole of Britain, and many others, I suspect. Father, an outspoken man, explained in no uncertain terms that we should expect the worst, and paid no attention to Mother's attempts to cover up war's potential horrors.

We all sat there munching celery and cheese; this was our current Sunday tea menu during the winter, together with those lovely thick slices of home-made bread and butter, followed by prunes and custard. Wide-eyed at Dad's account of war, we constantly fired questions back at him as three-, four- and five-year-olds do, when learning of something new: 'Who started it?' 'Will we have to help?' 'Will Hitler come to Asselby?' 'If they bomb our house, where will we live?' These were some of the typical questions.

During the war's early stages we saw little change, although preparations were under way, unbeknown to us. As the war progressed the radio news bulletins kept people abreast with its events. We dared not even turn the page of a book whilst the news was being broadcast. The contraption, mainly named a 'wireless', was not simply a mere matter of just plugging in and switching on. These basic stationary sets required an aerial, a battery which could weigh anything up to half a stone, plus a wet accumulator; the latter was the least trouble because a mobile service collected, recharged and returned them to the door at a cost of a few pence per head. The battery sometimes caught one out.

Mr Arthur Thompson was the village's 'wireless expert'. Everyone would call on him, should the old talking box refuse to function. Many times we knocked on his door requiring his services during his evening meal. He would open the door and explode into a flurry of crisp expletives which really bore no substance, as he would often be at our house even before we arrived back home, proving his dedication to his hobby.

Perhaps the norm is that never is a very long while, but I doubt if the sight which greatly excited us kids, and yet was a grim reminder for those who had family or relatives elsewhere in the war, will ever be repeated.

*Related Events*

It was in the early 1940s when a company of soldiers entered the village of Asselby on a 'manoeuvre'. They marched into the village headed by a band playing. Column after column of them appeared around East End Farm corner, whilst almost the whole of the main street was taken up. We marvelled at their beautiful immaculate uniforms and the precision of the columns. It was a truly magnificent sight, and none of us kids ever envisaged it. We got an even better view of them when they had to halt as a result of the railway gates being closed to road traffic whilst the train passed. They then marched to Barmby and crossed the River Ouse over the railway bridge, which again would require precise coordination.

The next time we had the pleasure of their company, it was even more exhilarating. On this occasion they invaded us, whilst there were not so many soldiers they virtually filled the village with tanks, guns, lorries, jeeps, etc., taking up residence in the paddocks, the farmyards and the lanes. Their headquarters and cookhouse were in the yard of Village Farm, the home of Mr Fred Johnson and family. They cut off branches from the trees and bushes to camouflage their vehicles. Gates and posts were demolished, and so on. However, this was military action, if only in practice form, so no one could complain. Barmby too had their share; they seemed to be dashing about all the day and night in their vehicles, and they in fact built a pontoon bridge over the River Ouse at Barmby.

Everyone helped those soldiers in some little way, but it was very difficult trying to ensure that no favouritism appeared to be taking place as there were so many of them. The womenfolk baked them titbits, and I remember it was the month of May, when my father's first outdoor salad crops were ready for sale.

Once they saw us harvesting and packing these fresh lettuce, radishes, spring onions, cress etc., that was it: as fast as

*Related Events*

us lads delivered one order, we were given another, until they finally cleared us out.

My father remarked that it was the fastest turnover he had experienced. 'Sold and money in the till within a few days,' he quipped. No doubt they had heard the sergeant major bellowing that an army marches on its stomach. Many times, however, substantial or not, those chaps greatly appreciated those salads and we were intrigued with their survival biscuits. Packets of the things were dumped all over the place, but we soon discovered one needed the teeth of a horse to eat them.

Just as mysteriously as they appeared, when we returned home from school one day we found much to our sorrow that they had disappeared, never to set foot in the place again. I often wonder, as these chaps came from all over England, if they remember exactly where they were camped up on manoeuvres; or were their minds fixed firmly on where they were ultimately destined for, hopefully to get the whole sordid business over with? Only they could tell, no doubt.

Another incident which I can clearly recall occurred on a lovely summer's evening, on 17 August 1943. A four-engine Halifax bomber, with a crew of seven and a full load of bombs on board, ran into difficulties close to the village of Asselby, and began to lose height rapidly, skimming the village rooftops, careering on extremely low over grass fields, slicing the top off a large ash tree, and finally doing a belly landing before coming to rest in a 40-acre grass field between Asselby and Newsholme. The aircraft was one of the JD 370 Squadron, based at Breighton RAF Camp, and was destined for Peenemunde. They had taken off at 2114 hours from the main W−E runway, and one can imagine the relief these chaps must have felt when coming to rest safely, and being able to climb out almost unscathed. The pilot had shown great skill in bringing the stricken plane down, and his task was said to have been aided by the collection of debris that built up underneath the aircraft's belly as it slid across the ground before coming to rest. First on the scene after the crash was a large herd of inquisitive bullocks. Wireless operator Sgt V McKinley was the only member of the crew to be slightly injured. The rest of the crew were namely Sgt A Scrivens, Sgt A J K Steven, Sgt A McKensie, Sgt F Cartwright, Sgt J Winchester and Sgt D G Cummings. I would say these men enjoyed a well-earned pint that evening back at the camp, and no doubt were recounting their lucky escape. I maintain it is because of these brave and outstanding men, and many thousands like them, that we are able to enjoy a flourishing Great Britain of today.

The next bunch of chaps wearing khaki uniforms who we had the pleasure of watching were no strangers to us. None other than Sergeant Abraham's 'Home Guard' mob displayed their humble accoutrements on Asselby's streets. Sergeant Abraham was a Goole man, and the son-in-law of Mr J Heron of Goole, the well-known millers and corn merchants. In fact they all came to live at Asselby after Goole was bombed. Mr Abraham was none too popular at times in the eyes of his men because he, being an office man, had far more spare time than they had. Therefore he would look forward to the practice and drill sessions in the evenings on Sunday and other times with much more enthusiasm than they, who had had a long hard day's graft on the farms. However, Mr Abraham was not without a sense of humour and nor were his men.

They sometimes had mock battles with other village Home Guard mobs. Other times it would be fire drill. On one such occasion Knedlington was supposed to be expecting an invasion. Therefore they were positioned all over the area. Ira Hutton and Ernest Dennison were ordered to cover Yarmshire railway crossings. They had been hidden in the dyke all the night, when at about 5.30 a.m. they heard this pitter-pattering sound approaching in the darkness.

'They are here!' exclaimed Ernest, to his bewildered colleague. They looked eagerly at each other and one of them said, 'Aren't we supposed to halt them?'

At that they sprung into the middle of the road with fixed bayonets and shouted, 'Halt! Who goes there − friend or foe?'

But alas, the suspect enemy still approached in silence, and seconds later a herd of cows were staring them in the eye, much to their relief and to the amusement of the cows' owner, Mr Stead. He merely remarked, 'If that's all you chaps have got to do, come along with me! I'll find you some work!'

Ira was a tall man who needed the largest footwear one could buy, which together with his flat-footed gait made marching difficult for him. Sergeant Abraham would snap, 'You are out of step, Hutton!' To this, Ira would dryly reply, 'No sir, it's the others!'

Ira was an excellent horseman and a competent farm man. He and I became good friends when I started work in the village; he taught me many things pertaining to village life and how to survive on a meagre income. He always killed two pigs, kept a pen of hens, cropped his garden continuously − all of which cost him very little, because he was quite good at scrounging and wheeling and dealing with the farmers.

He would be up at the crack of dawn collecting coal from the railway which had fallen from the wagons. The plate layers always threw it down the embankment as they maintained the track anyway, therefore Ira insisted he wasn't stealing it. His wife was a jolly woman, and they had three young boys who would have brought a ray of sunshine into anyone's life. I used to go to their house in the evening, and we would sit drinking tea, laughing and talking about any mortal issue. It was on one of these occasions when Ira told me of their confrontation with the cows. Dolly often poked fun at his Christian name − Ira − and between us we would try to envisage where it might have been derived from. Today it would not be so difficult. No one could have possibly foreseen the sad and sudden tragedy which was to shatter their humble, perhaps mundane, but happy lifestyle. It seemed so incredible that their second son Keith should be killed on the railway line, the very place near where the lad had grown up. He had been strictly taught how to use it by his father, and most of us had to cross the line at one point or another.

Keith unfortunately disobeyed one of the railway company's rules, which was in fact a finable offence, and was clearly stated as such on each gate. He left open the occupational crossing gates and drove his tractor and trailer to and fro across the line throughout the whole day. Perhaps this wasn't such a perilous practice, given good light and 100 per cent concentration from the offender. However, when Keith drove his tractor across the line and into the path of a giant railway engine travelling up to 60 mph, it was about five o'clock on a November day and therefore twilight, and rain had started to come down, causing him to wear a large overcoat. The tractors had no cabs at all in those days; therefore a greatcoat was essential when one was sat on a tractor seat open to all the elements. It was suspected at Keith's inquest that the large, upturned collar of his coat restricted his vision and contributed largely to his accident. Sadly, by his overfamiliarity, he paid the price with his life. It was a very sad time for the family, and indeed for the whole village, to lose such a bright eighteen-year-old lad who inspired everyone. I actually heard the impact of the accident, not knowing anything about it at the time, of course. Perhaps it was of some little consolation for them to find that Keith's body was still intact − unlike his tractor, which they picked up from the railway lines in pieces small enough to be manhandled.

On another occasion we came out of chapel at the end of our Sunday morning session to find Sergeant Abraham putting his men through rifle drill. Apparently their performance was not up to his expectations, for they were going through the same moves, resulting in Walter Davy's patience running out, it was 12.30 p.m. and no doubt Walter's mind had succumbed to the aroma of roast beef and Yorkshire pudding, for he suddenly bellowed, 'Don't you have dinner on a Sunday, Sarge?'

*Related Events*

'Yes, I do,' replied Sergeant Abraham. 'Why do you ask, Davy?'

'I would have thought that would have been obvious to you, sir,' said Walter.

'Whatever we do here,' snapped the sergeant, 'must be regarded as if we were in the front line — and that goes for you all!' But he did compromise, shortly afterwards he dismissed them and the rest of the day was theirs, as they say.

Despite its potential dangers, that beautiful piece of technology, the Hull and Barnsley Railway was a majestic feature of Asselby in those days. Everyone knew it had to be used with respect, Keith was not the only one to occasionally let familiarity breed contempt. Generally speaking, fog or very bad visibility were about the only conditions that affected the basic rules when using the occupational crossings, especially with cattle, or indeed any animals, a task which I did almost daily when employed by the Heseltines. During a thick fog and with signals obscured, it was down to listening, which again could be misleading. However, the railway's fog signalling system gave some assistance, as the cracker could be distinctly heard and the engine driver proceeded only on hearing its report. Under the former abnormal circumstances, two or even three persons would coordinate and get the animals across the line quickly. The railway was the cause of many a mother's anxiety, as the fascination for these giant steam trains roaring close to the village drew the youngsters like a magnet.

Occasionally we got an extra treat, being able to get a really close look at the jet-black monsters, when the bridge over the River Ouse at Barmby was open for ships to pass through. The trains then had to stop. The engine driver's favourite trick was to allow us to stand close to the engine, then suddenly release steam, which even if one was expecting it literally would frighten us to death.

One morning we were picking potatoes when a train was held up, and the driver rolled us several large lumps of coal

*Related Events*

from the tender in exchange for a sack of potatoes. We were amazed how quickly they roasted us some potatoes on the fireman's shovel! As for the coal, I suspect they deliberately gave us the largest lumps they could in order to have a laugh at three or four of us struggling to lift them.

We spent hours playing around the railway. The culverts and the small section of fire dykes were usually dry, but they also served as ideal habitats for eels, pike and newts when carrying water, which we used to catch by very primitive methods. Providing the plate layers or the railway police were not around, the signal box men, too, kept a close watch through their binoculars for trespassers or any foreign objects on the track. All the gates which the public used were equipped with stout chains, enabling them to be locked as well as clasp-fastened if the users chose.

Again, the daily and nightly sights and sounds of the railway's activities were regarded as normal rural ones which the village people lived cheek by jowl with, and alas they were not subject to complaints, as are the same ones of today.

It beggars belief and is downright nonsense when such petty noise-related issues appear in our courts. One wonders why on earth they all have not something better to do. Such a waste of time and money! I refer to one such case that I read of in the press where a case was brought before the court regarding a noisy cockerel's early morning serenade.

This is about as pathetic as taking out a summons against a hotelier, or a guest house landlady, because of the early morning serenading seagulls, which I am sure most of us have had the pleasure of hearing whilst lying in our seaside holiday beds.

I can only say had the person in question been around when we used to lie in our beds on a winter's morning and listen to the concerto of some twenty cockerels, he or she would have committed hara-kiri. Having been in our beds since 8 p.m., we would be awake and ready to get up by

5.30 a.m. The Everatts, Claytons, Lapishes, Ellwoods, Huttons and ourselves all had more than one cockerel on board, which were all within earshot of our house. We would even have little bets on whose cockerel would set up the crowing. As with people's singing and speaking, most cockerels have a slightly different tone, style, tempo and crescendo as they perform their daily early morning ritual, which we learned off by heart.

Those birds somehow were part of the community; no one regarded them as a nuisance. The trains too, lumbering on the outskirts of the village during the still of night and early morning, had a special effect. They vibrated the old sash windows and gently jingled the pots in the sideboard cabinet. The clank of the wagon's wheels passing over each expansion space in the line sections became more faint as the trains went on their journeys. These sounds and noises one could dream through and regard as normal as breathing the air.

The sound we all dreaded, however, was that of the air raid warning siren. It was so menacing. That high-pitched wail across the still of the night spelled out danger emphatically, and people reacted to it in various ways. There was no special code of conduct during the time between the warning and the all-clear blasts of the siren, but absolute blackout was the ultimate rule. Our neighbours, the Palmers, being two ladies on their own, were very frightened during a raid; they would take refuge in an old cupboard. My father would always get up if the raid occurred whilst we were all in bed. The very youngest children usually slept through the hubbub, the rest of us would be catnapping. The wardens' voices could be heard from the street below. Sometimes my father would have a few words with them from the bedroom window, which overlooked the street at one side.

Of course we knew nothing compared to those people living in the towns and cities, but even so the threat was there. The blasts from the bombs being dropped around Hull continuously rattled the doors and windows during a raid, which spoke volumes of what it must have been like there. Mother was very frightened on those occasions, perhaps mainly for us children.

The nearest bombs to us fell at Newsholme, and badly damaged the Selby Road, as the crow flies, only a mile or so from us. Goole was also bombed; I suspect that the docks would be the intended target. I clearly recall Christmas morning 1943 when one of Hitler's flying bombs gave us a visit. It was about 5 a.m., when we were awoken by this loud and obviously very near sound of an engine, which appeared to be travelling slow and laboriously. My eldest brother, Alan, called out to Dad from his bed and said it was Ernest Newham arriving into the village with his steam threshing set. The machine, with its very large iron wheels, did make an almighty din on most occasions.

By this time we were all wide awake, but the last thing on our minds was what Santa Claus had brought us. 'That's no steam engine!' retorted Dad, in his loud clear voice. 'It's some bloody mischief of Hitler!'

Whatever the beast was, it passed over the house tops perilously near, before its dreadful sound had faded away and we had even contemplated looking at our presents. It was inevitable that the mysterious intruder had a mate, which could soon be heard approaching. At this we all dashed to the windows and actually faintly saw the thing on the grey morning skyline. The black ghostly figure's stealthiness was quite menacing; when it disappeared from our sight and its odd-sounding engines faded away, we could only regard ourselves lucky that it had done so. 'There may be others,' said Mother anxiously.

'If there are, so be it,' replied Dad, 'for no matter what precaution we take, we cannot defeat the mechanism of those wretched things!'

Shortly afterwards, Dad had a roaring fire going, and

*Related Events*

Christmas began and proceeded with no further dramas, thank the Lord. It was some consolation to know that the Germans were not having it all their own way. We used to watch huge squadrons of Halifax bombers flying over the village on their destination for bombing raids over Germany.

When the war officials issued us all with a gas mask, many people regarded them with contempt, perhaps mainly because they were not sure how reliable or effective they would be. One was not supposed to be more than a few minutes' retrievable distance away from one's gas mask. Therefore we had the burden of carrying them to and from school each day. The original ones were prototypes, so whether or not they would have been life-savers in a crisis no one could tell.

We were given weekly drill sessions with them, during which time we discovered that the major fault in them was the steaming up of the perspex mini-window. This again was not perhaps a fault, but more a case of whether or not they would save lives.

However, the powers that be made several modifications to the tiresome contrivances, which we fortunately were never required to wear for real. It was difficult, especially in the case of babies and the very young children, who were literally in awe of them; so much so, some mothers used them as a behaviour inducer at times.

The war provided a bonus towards us lads' pocket money in some ways, although it was no handout. I can recall one summer when there was a plague of cabbage white butterflies. Cloud upon cloud of the things emerged as if from nowhere. The Ministry of Agriculture in desperation came up with the most primitive method of eradication, which today would be regarded as a waste of time. However, they alerted the rural schools' headmasters and instructed them to encourage as many as possible of us schoolchildren to join in the campaign to combat the little white intruders by killing as many as one could, at a reward of one penny for one hundred.

A four- or five-foot stick with a square of ply board nailed on the end of it was to be our weapon of destruction. The clover fields seemed to be the butterflies' favourite habitat. Literally millions of them formed a dithering white blanket over the countryside.

I suppose to the layman we looked something like ballet dancers in despair as we jumped and leapt about, swatting the defenceless creatures. Of course they had to be retrieved and presented to Mr Walker each morning, who had the job of checking our count and putting up the money. We used to work in pairs, one killing and the other counting and picking up. We would spend about an hour each night or stay while we had bagged 600 each to make a round tanner each before going off to something a little more exciting.

Now I cannot pretend to know whether or not the people who administered this campaign against the white butterfly monitored any statistical results. I would say generally speaking it would be very difficult to obtain a valid result. However, as far as my personal observation goes, I can honestly say I have never seen a repetition of that magnitude of that particular pest ever since.

Our next crusade was the collection of wild rose hips; the Minister of Food no doubt had something up his sleeve for these. They were a bit better as a paying proposition than the butterflies, but no easier task, although there was a heavy crop of them that particular year. Whilst gathering them, we often had to compete with nettles, thistles, plus the sharp thorns on the wild rose bushes. One stone in weight was the minimum that Mr Walker would weigh in, although it wasn't the minimum weight that concerned us, it was the maximum. When it came to carrying two stone or so of rose hips to school, plus one's gas mask, coat, packed lunch etc., it became something of a struggle.

But there was that scarce commodity called money in it for us, so the incentive was there. Mr Walker had been instructed to consign them to their destination in 1 cwt. sacks via Barmby Station, and I seem to recall that he was responsible for getting them there too. But when he had a few hundredweight for despatch, he would grunt, 'Well, lads, it's your heyday, not mine!' So he would detail two of us in turn to place a sack of the hips into a barrow and wheel them to Barmby Station — not that it caused any detriment to us. On the contrary, it was a break from the drudgery of lessons, and there was always plenty of activity in the station yard to interest us. Also, we experienced the dry sense of humour of old Bob Lofthouse, the stationmaster, and his unprintable comments regarding our mission!

As the war brought about the rationing of just about all and sundry, the services of the village tradesmen were greatly appreciated. Mr Wheelhouse was the local tailor, and he lived in the end cottage on Pond Hill at Barmby. He would make, and more frequently repair, trousers, coats, shirts, etc. for a very affordable and somewhat ingenuous fee.

So when Mother had a number of garments which required attending to, in order to obtain the maximum wear from them, we would, during the school's lunch hour, pay Mr Wheelhouse a visit. He was a small Ronnie Corbett type of man, with an iron calliper on one leg. He was always so cheerful and welcomed us into his home with open arms. He had a grand piano which he played beautifully, and his love of it superseded the purpose of our visit. He would hobble over to his treasured piece and play us an introductory tune. He would then enquire which songs we were learning at school. As he kept abreast with the music and songs of the day, he soon had us all in accompaniment with him. Often he had a couple of small pans on his open fire with some potatoes cooking in them. They too were neglected in preference to entertaining us kids. When his clock said 12.55 p.m. he would quickly run through the work we had brought him, give us a date for collection of the same, then hurry us through the door with a few anxious words of encouragement that we made it back to school by one o'clock.

A larger amount of his work was provided by the farmers, the horsemen and the farm labourers. They wore, essentially for the nature of their work, thick serviceable breeches, leggings, jackets and such through the winter months, i.e. corduroys, Derby and cavalry twills, which Mr Wheelhouse made and kept in repair for them. But even so, the revenue from these customers being subject to credit, rendered our small cash jobs more advantageous for him.

Mr Jack Hill of Knedlington, alongside his job as a farm foreman, performed the role of the local cobbler. He would work late into the night, tired after a hard day's slog, in order to oblige his customers — or perhaps, to be more precise, to eke out his meagre pay; for he too, like many of us, had a large family to support.

He would quip when us kids handed over the few bob on picking up the repaired footwear. 'Don't bring them back! Not before you have worn them, anyway!' But alas, there were times when he had to reluctantly inform us that the task was impossible or at least not worth our money nor his time being spent on them. I would concede that white and hard winters were the norm of yesteryear, and these in themselves took their toll on our footwear.

Whilst our elders could afford skates of some description, we had to make do with strong nailed boots. Ice-skating was the main winter sport for country folk in our younger days. These days, not due to lack of rainfall, but largely down to flood prevention schemes, and much deeper and better-controlled drainage systems, we neither have the water nor the ice — certainly never both at the same time, an essential ingredient for ice-skating.

Asselby and Barmby marshes, and indeed a large acreage around Howden, flooded annually. Keen and severe frosts nearly always obliged. The marsh floods provided ideal skating sites, mainly because they were relatively shallow waters and therefore safe in a relative kind of way. Games on the ice were great fun. On a Saturday afternoon and Sundays the grown-ups would have ice hockey matches. Their skates were nothing flash. The farming fraternity wore very strong hobnailed boots, essential for the nature of their work, which were also a good combination for the screw-in and strap type of skate they used. The high tops of the boots I refer to provided excellent ankle protection to the wearer, and indeed from the stout old sticks they cut from the bushes.

Those Sunday skating sessions passed with gaiety and grandiosity. Even the very senior citizens, if the walking conditions were favourable, would don their thick coats and scarves and hobble down to the marsh for a glimpse of the fun. I often noticed tears in the eyes of some of them, perhaps due to the nip in the air or maybe from reminiscence of the days when their now inflexible limbs allowed them to do likewise. The old men would argue which previous year or winter provided the longest skating season during the day, and so on.

I can recall the following splendid people being on the ice together: John Palmer, Cyril Bolden, Harry Barker, Walter Davy, Frank Andrew, Harry Coatsworth, Fred Harrison, John Pearce, Eric and Clifford Ellwood, Maurice and Ernest Hutton, Bill and Dick Clayton, Charlie, John and Noel Stead, Norman and Charlie Winter, Robert and Margaret Sales, Peggy Earnshaw, Audrey Bolden, Kathleen Clayton, Harry and Mr Seefield, Dorothy Ellwood, Bill, Jack and Ivy Walker, Leslie Stewart, Ernest Dennison, Tom Bristow, Alan Clark, my eldest brother, plus us younger relatives, brothers and sisters of the former.

At 3.30 p.m. the party would reluctantly begin to dwindle. The farmers and their men would be the first to leave, having horses and other livestock to tend. They very rarely returned to the ice in the evening, as most of them would head for the towns. But given a moonlit night we kids were allowed out on the ice on a Sunday evening. It was absolutely elating being out on the marsh under a full moon. The calls of the coots and moorhens and the screeching of the old barn owl searching for his supper seemed so clear and near in the silent thin air. On a few occasions we saw Mr Reynard on the hunt for his supper too; his piercing eyes would shine brightly as he passed through the shadows of the bushes around the unflooded perimeter of the marsh.

Of course there was nothing to fear out there, and our parents were at ease knowing that the ice was safe as the severe frosts prevailed. However, we took a fright one night. While a large crowd of us played merrily, one by one we became aware of a tall figure standing motionless and staring at us. As the word got round we decided to scarper, and I can honestly say that Sebastian Coe would not have had a look-in; we did not stop or look back until we reached the railway gates which we cleared in turn like the Grand National contenders, while our mysterious figure stayed put.

We never discovered any explanation or clues to this incident, but I can truthfully say it was no illusion. We had no suspicions about who it might have been. This was rather odd, because in a close-knit community, as these small farming villages were in those days, everybody knew each other, so to speak, as regards movements, habits, temperament, capabilities, sense of humour and so on.

We had, in fact, some frights and pranks played on us by the farm horsemen, who were always about in the farmyards and stables on the dark winter evenings; but on those occasions we always discovered who our tormentors were. 'Fox and hounds' was our favourite winter game, which was perhaps a sophisticated version of hide-and-seek. With all the yards full of hay and corn stacks, and no street lighting etc., there was no shortage of hiding places. The game was a great time consumer. Often our mothers were calling us in even before the first game was up.

*Related Events*

We always welcomed the arrival of the steam threshing set into the village, especially if it coincided with our school holidays. Not only did it provide the fun chasing after the rats and mice, but there were always the light jobs us strong lads were given the opportunity to do on the threshing days so as to earn a few bob. None of the farmers seemed to own a hose-pipe, and the nearest tap could be fifty yards or more away from the engine. We would think we were doing well carrying half buckets of water to the forty-gallon tank placed near the engine when it became almost full. But it quickly became a disincentive when we returned to find the tank empty. Old Harry had lowered his suction pipe into the tank, and like a greedy elephant sucked up the lot into his engine's boiler. He would chuckle when we returned with more water and say something like, 'Didn't you notice that big hole in your tank?'

Harry and Ernest Newham of Howden toured the area, contract threshing for the farmers. For all that they were father and son, animosity often erupted between them. Harry, the father, had a dry sense of humour, but he could also be cantankerous. This was highlighted on the dark winter's nights when, if they had a long move, it could be six o'clock or later by the time they arrived in the village.

Having been up since five o'clock in the morning, one could understand their plight, bearing in mind the lighting they had was meagre; paraffin oil lamps were all they had to rely on. Many of the farm entrances were awkward enough in the daylight, let alone darkness, and sometimes snow and ice packed the yards and drives etc. The overall length of the set, plus a restricted quarter-lock steering mechanism, was a handicap too, even under good circumstances. The threshing machine would have to be shunted into certain yards with as many men as available hand-steering it at the other end.

These were the times when the sparks would fly, and I don't mean from the engine's firebox. Harry and his son would curse each other black and blue, each claiming they could handle the engine better than the other. On one or two occasions, such was his anger that Ernest mounted his cycle and left his father and the farm men to it. Of course it wasn't always like that. Indeed, it was in their best interest to ensure that it wasn't. After all, it was their living and their ultimate aim was to be as obliging and as amicable to their customers as possible.

The by-products from the threshing day, i.e. chaff, pulse and straw, were not, as is the case today, regarded as a nuisance and a hindrance to field procedure. In fact they were regarded almost as an equal to the grain, and therefore stored on threshing days and used as a supplement to the diet of the in wintered livestock. Even though it took two or three of us youngsters, as opposed to one man or woman, we were often given the job of carrying chaff and pulse to the store, which unlike the grain was not subject to any scrutiny.

The bright winter sunshine made for a better threshing day scene; those beautiful black and copper coloured steam engines held a kind of attraction to one and all. They functioned so silent and smooth, and their owners tended them with almost as much care as they'd give a loved one.

When the team took a tea break, the engine man would be the last to leave the scene and the first to return. We loved to stand close to the gentle giant. With all the driving power shut off, it gave off a unique heat as it stood silent except for the quiet hissing from the steam valves. A mixture of steam, burnt oil and burning coal smoke hovered round it. However, once the internal combustion engine came on the scene, economies superseded sentiment, sadly resulting in the steam era being here today and gone tomorrow.

So the old threshing machine acquired a new counterpart, so to speak, although already its redundancy was looming, with the prototype combine harvesters beginning to appear around the mid-Forties. But even so, the threshing machine and its new partner, the tractor, survived for another decade or so and

I am afraid it has to be said that the tractor proved to be more serviceable and indeed a more versatile part of the threshing unit. All too late, however, as regards helping the inevitable fate of the same. We did in fact have a threshing contractor in Asselby during the Thirties. The Taylors of School House ran two steam threshing sets, although they were none too popular locally, perhaps mainly because they seemed to concentrate largely on their Goole area customers and therefore were not available when the locals required them.

It was actually no secret that one or two local farmers regarded the Taylors as a little slow. Now, whether or not there was anything in their view would be neither here nor there, for I would think it could only be down to a very minor detail anyway. Generally speaking, however, the rule of thumb was that a farmer would expect a 6-yard by 4-yard stack of corn, or the same size Dutch barn bay of corn, to be put through the threshing machine in half a day, 4½ hours; or a 9-yard by 5-yard stack of corn in a whole day (approximately 9 hours) — being of course subject to one or two slight discrepancies either way. I don't know; perhaps the Taylors were not actually achieving this. But I do recall some controversy regarding the same issue when I started work, and became actively involved in the threshing operation. By this time, Clifford Blacker of Wresstle had taken over from the Newmans as the area's threshing contractor.

A chap called Stan Parkin managed one of his sets. Bearing in mind that this was now the mid-Fifties, times and men's rights and priorities were changing and commanding more recognition. Stan was a time man. Prior to him, if a day's threshing ran fifteen or twenty minutes or so over the time, especially if it was a special consignment of grain in, it was accepted and regarded as normal practice by all. But alas, Stan would not wear this. He would chunter on about not being paid overtime and how such circumstances had made him late home, having to clean down, pack up and move to his next

*Related Events*

farm, etc. I suppose he had a point, but his attitude greatly annoyed the farmers.

However, as I say, changes were afoot in many departments, and therefore Stan's mischief eventually became unacceptable. I feel I must add that it was a sad day when the combine harvester finally pushed the threshing machine onto the headland, for not only was it redundant but so also were the odd-job men who made a living through the winter months following the machine on its tasks. There were many of these rough characters around.

One outstanding one was old Bill Walton of Howden. Bill was a large, dark-haired man with the features of a giant. While a little slow, he had the strength of a horse. He was absolutely unflappable, with not a care in the wide world.

One cannot help but envy the positive and laid-back lifestyle of these characters of yesterday. As Bill often boasted, he had no overheads at all. He would quip, 'What Bill earns goes into Bill's bank!' — patting his back pocket to confirm his explanation. He always wore two pairs of trousers, which raised the eyebrows of those who did not know him when he would partly take down the top pair to get to his so-called 'bank' when in a shop, chippy, pub or whatever; and I believe that the bank was always in good funds.

During hard weather, the likes of Bill would often stay the night at the farm where they happened to be threshing, and so avoid a cold ride home and back again the next morning. After perhaps a few pints in the local, they would bed down in the farm buildings, which were always well stocked in those days. Dairy cows, for example, were tethered in stalls in the cowshed. There was no better place to be on a cold winter's night. The cowmen went to great lengths to ensure their cows were comfortable, the windows and doors were draughtproofed with straw and old sacks etc. A pile of hay in the corner was all one needed for a warm night's sleep.

Again these men were very tough. They would wake the

next morning, often break the ice on the fold yard cattle trough, wash in it, dry off on a hessian potato sack, and they were ready for work. The farmers would always supply a good rough sandwich; in return they would help to feed the midwinter stock before the threshing day began.

Harry Moffat was another regular face, and many a farmer claimed that the threshing team was incomplete without him, for he was an expert straw stack builder. This attribute kept him in high esteem, as it was of paramount importance that the straw stack was built, like any other roof, to keep out the rain and snow. I am afraid that a great number of men could not ensure this, but Harry could.

The secret of this particular job lay in one's ability to envisage from the corn stack about to be threshed what the dimensions of the straw stack to be made from it were to be. Again, old Harry excelled at this. He would keep his eye on the progress of the threshing, and by doing so he seemed to know exactly when to roof up his straw stack. Very often the last lot of straw to leave the threshing machine completed his stack. While these chaps were regarded as mere labourers, in there field I would say they were worthy of much better recognition.

The rats and mice could play havoc with the corn stacks, especially the ones left until the warmer weather of the early spring. The Ministry of Agriculture stipulated that the area around stacks showing signs of infestation should be surrounded by wire netting. The rats made excellent catapult targets as they desperately searched for an escape route along the wire netting. Most lads had a catapult. Merv Walker was a crack marksman with his catapult. Providing the dogs were not in the way, he could lay out the rats as fast as they came out.

We spent hours in the summertime sat behind the may bushes on the marsh with our catapults, waiting for the wood pigeons to settle on the tall buttercups in seed. They were very partial to them. The feral pigeon too, which was regarded as a pest in some respects, provided us with some fun and at times frustration. Some early nineteenth-century architectural plans of farm buildings included a brick-built pigeon cote. The feral or domesticated pigeons would congregate in there – hundreds of them. We spent endless time building sheds and converting our lofts and false roofs into cotes, in the hope that we could have our own flock of pigeons. But it was all in vain. On release, they always made for Tommy Ellwood's cote, which was the most populated one in Asselby. We would approach Tommy in his garden and inform him of our plight. There would be perhaps about a hundred pigeons on his cote roof, three-quarters of them Blue Bars. He would chuckle and say, 'Which are yours, the blue 'uns?' He'd continue to taunt us by saying, 'Get up there and catch them, then I am not hindering you!'

Of course, we never got them back, but it did not deter us from trying again. We used to fire small potatoes at them with our catapults, keeping them on the move, and hoping they just might come home. Alas not.

The situation was rather similar to much earlier times when the lords of the manor monopolised the domesticated pigeon. Only they were allowed to hold the large cotes, which in those days provided meat, and the eggs, too, were marketed. Even in the 1930s, the times which I refer to, the pigeons were marketed and eaten. The dung, pigeon droppings, was reputed to be rich in nitrogen properties and therefore much sought-after by the allotment holders and exhibition vegetable growers. But again, the manor houses had beautiful walled gardens that were utilised to the full by employed gardeners, therefore acquiring the pigeon manure would not be easy. The most unjust aspect perhaps of these winged intruders was the fact that they were scavengers, feeding on anyone's crops and in anyone's yards. Their beady eyes might focus, to add insult to injury, on many of these people who were also paying rent to the squire, making it difficult, if not impossible, to complain about their depredation.

So much for the humble pigeon and its past values, which like many of the rural traditions are no longer viable in our modern society. The Thompsons of Asselby Mill also fostered a large cote of feral pigeons. Mr and Mrs Thompson were the last people in Asselby to use a horse and trap as their mode of transport. Mariah Thompson was a stately-looking old lady with a heart of gold. If we were playing in the village or taking the younger members of the family for a walk in their pram along Knedlington Road and we met, they would stop the trap and Mariah would step down from her seat and give us some sweets. She would always have a look at the baby in our charge, warn us to be careful with it, and not go too far away from home — not that we hadn't already been told.

The horse would be fidgeting and stamping its feet in boredom, wanting to be on its way, only old Bob's firm grip on the reins preventing it from doing so. Then he would hold out his hand for Mariah's grip as she climbed back on board. The horse would lunge forwards as soon as his rein was free, and away they went.

Mariah often visited our home, and again her kind and good nature was reflected in the fact that she always brought us something. Quite often it would be something we had plenty of; but as the saying goes, it's the thought that counts, and anyway with a large family, as we were, one could never have too much fare. Sometimes she would bring a large pigeon pie — her very own recipe, she would claim — which was delicious.

She would sit and pass on her culinary expertise to my mother, and she was also well read in the field of the ills and chills of children. Even though most of these old ladies of the 1930s were good-natured and gently spoken, many very young children were afraid of them initially — including myself, I must add.

Perhaps it was their somewhat weird dress, for there was no competition in the field of fashion; 'uniform' would perhaps be a more apt word, for they all wore long black skirts down to their laced black boots, a black bodice or blouse, a black hat and shawl. They very rarely visited the dentist, and their weathered and wrinkled faces had never seen such luxuries as moisturisers, rouge or the like.

Admittedly, none of the former create personalities, temperament or mannerisms etc., but meeting one of these old ladies in a bad light, so to speak, could be an experience for a child. Mariah would not thank me were she here today for mentioning her neighbour, Mary Lambert of Waterworks Cottage, for they had only contemptuous feelings towards each other.

Although no one could deny that Mary was a loud, brash, brazen-faced, outspoken woman, she was basically quite innocuous with it all, and like Mariah she was fond of and good-hearted to children. We could never pass her house unseen by her, but in any event one always heard her before one saw her. She was often referred to as the 'Howden Gazette', for anything she did not know as regards local news wasn't worth knowing; but again her mannerisms bore no malicious intentions.

Mary was yet another hard-working, flexible, much sought-after woman. During this era there were many of her type, for though they were never employed on a permanent basis they were never out of work. Their mundane lifestyle held nothing to be desired. A person similarly employed today of course is regarded as self-employed, but the former was so different, so much more fair and just, unlike the latter, which is subject to government legislation and false inducements and pitfalls.

Therefore, mundane or not, in my book the former were happier, more content and even more secure people, who would not have liked today's version of their status one little bit. Mary saved and managed her money and eventually bought and paid cash for her cottage without the intervention of a bank or mortgage millstone — that in itself made for the better one's peace of mind.

The name of Mary's cottage, in fact one of a pair, signified its relevance to the waterworks, and in fact formerly served as staff cottages when in 1901 Thomas Sinclair Clarke Esq., Lord of the Manor of Knedlington, sank a borehole at Asselby and erected the waterworks, which stood only a few metres from the cottages. The project subsequently supplied the whole estate of farms and cottages with piped water. So prolific was the source of his supply that the squire offered to supply Howden with piped water too, as it was without it at the time; but negotiations failed and it never came to pass.

Thomas Sinclair Clarke Esq., MA, JP, was Lord of the Manor of Knedlington, and the principal landowner at Asselby, Barmby and Knedlington, along with Lord Leconfield, Revd T W Brooke, Mr J W Shaw, and messrs Hammond. The Clarkes owned their land from the mid-1700s. Knedlington Manor, the estate's headquarters, was built circa 1841. The Old Hall was the former residence of the Clarke family during the latter part of the sixteenth century. The Arlush family owned the estate, which passed from them to the Terricks.

The Bishop of London, Dr R Terrick, was born at Knedlington and died in 1777. He divided his time between London's commerce and his landed interests at Knedlington. Thomas Sinclair Clarke, the son of Sarah and Thomas Rudd Clarke, Barrister-at-law, Knedlington, was born on 5 February 1831, and remained there with the winding-up of their enterprise. Mr Mortimer then succeeded them for a short term, and finally and to this day what remains of the estate still belongs to the executors of the Yarborough family.

The Clarkes were the most popular and widely referred to family. They put a lot of effort into the managing and maintenance of the estate, and took the greatest interest in the tenants on the estate. However, it appears that beyond their boundary and their circle of friends and tenants, things were not so rosy. Howden Town Council, for instance, held some contempt for the Seignior. No doubt it would be down to the goings-on in the Chambers of Commerce, or some sort of political or religious indifference. Who knows?

Does not the former suggest, however, that the saga concerning the disappearance of the family's historical monument and landmark have some connection? I refer to the monument that Thomas Rudd Clarke had erected on his land at Knedlington Corner in 1901. It was a beautiful concrete monument, complete with a supply of fresh water for both people and cattle, comprising a fountain and a cattle drinking trough, which I used myself on many occasions. It also commemorated his son, Thomas Sinclair Clarke, and bore their family crest and motto: 'The time will come.' The fate of the monument, albeit lost in mystery, would appear to be largely down to some misdeed, therefore I hold strong misgivings regarding the legality of the issue. According to an article which appeared in the *Times and Chronicle* on Tuesday, 11 July 1991, workmen from the council came with lump hammers, broke up the monument, and carted it away for infill. Now, if this was the case, and bearing in mind that the workmen were merely obeying orders, surely the onus lies on the shoulders of some council official.

It goes on to say that the treachery took place during the 1960s. Why, I wonder, as I suspect many people do, was the case not investigated at the time? Instead, there was a lapse of some twenty-five years before Howden Parish Council reopened the issue, proposing to replace the monument in some form. I see no reason for anyone to doubt the feasibility of the council's explanation, and when one looks at the undeveloped and undisturbed site of the monument, all is revealed.

The Yarboroughs, to whom I previously referred, were very influential people and fine agriculturalists. Their vast tract of land comprised some 30,000 acres in North Lincolnshire, dwarfing the Knedlington Estate by contrast. I would not go so far as to say this was the reason they purchased Knedlington, but I do know it was a good hunt run of open arable fields with virtually no fences or other obstacles, stretching from Boothferry to Barmby on the Marsh. I can recall those Lincolnshire huntsmen having some good chases over the former in the 1930s. They greatly enjoyed the vastly different countryside to that of their own.

The Yarboroughs (the Pelham family) owned their own distinguished pack of foxhounds and bred fine, carefully selected bloodlines of hunting horses. The 1st Baron Yarborough in fact transformed other very large tracts of land on the North Lincolnshire Wolds into prime agricultural land from a prairie of gorse bushes, bracken and rabbit warrens, much of it in fact land that no one wanted, even as a gift, until these fine agriculturalists set about conquering the wild, in about 1750. They stuck it through thick and thin and held the theory that bad times for tenant farmers meant bad times for good landlords. Whether or not the fact that land could be bought for £3 per acre, and the current rent hovered around 3 shillings per acre per annum, made good arithmetic remained to be seen.

In 1895 farming was once again plunged into a depressed state. This was when Lord Yarborough practised what he preached: when his tenant farmers began to fail, he himself came down with them, by greatly reducing his hunting costs and other non-essentials too.

The 1st Baron Yarborough was created in 1794; his elder son became the first Earl of Yarborough in 1837; at least three more earls of Yarborough followed on. The 4th Earl of Yarborough's title was, in full, 'Charles, PC, KG, Lord Lieutenant of Lincolnshire'.

The Countess of Yarborough, a very beautiful and stately lady, also in 1903 became historically unique in acquiring the titles of the Barony of Fauconberg and the Barony of Conyers, making three titles of great esteem at the time. Britain's heritage will retain the hallmark left on it by these fine enterprising pioneers until the end of time.

## The Changing Face of Agriculture

I suspect we all look back on life at some stage and draw the conclusion that if only I had done this or that, or not done this or that... Likewise, I suspect many of us find it hard to believe certain outcomes or certain aspects of our lives.

The one outstanding fact I found hard to believe and come to terms with was the fact that the day had come when I could not, after being made redundant, get back into agriculture after spending thirty-two years employed in the industry and fifteen years or so growing up amongst it; albeit I learned that I was not alone in the dilemma by any means.

When I started out in 1948 I was approached by at least four employers sincerely enquiring about my interests and intentions and offering full-time employment. This in itself was an enormous ego builder; the feeling of being wanted gave one immediate confidence. Furthermore, if one did come to grief with a post it usually took little more than a stroll along the street and a personal approach with a few exchanged words to become re-employed. That is, in fact, how I acquired the situations which I held, and can count them on one hand, during the thirty-two years I was employed in agriculture.

But alas, in 1980 through being made redundant I was thrust onto the job market, and was to discover that the whole system had changed beyond belief. Gone was the former simple, palaver-free initial interview, which again was usually settled one way or the other there and then. In had come the formal pre-arranged interview, preceded by an application form and the request for a reference. I could never see the point of the latter; the fact that an applicant had spent ten years working in his present employer's post would tell me all I

## The Changing Face of Agriculture

required to know of him, as in my case. For let's face it, there is no farmer in England who would keep a man so long had he any misgivings about him, large or small.

Often nothing at all would be discussed during one's initial enquiry about a post. A casual approach had become out of the question, and I was utterly astounded by it all. But that was not the worst of it, for some of the interviews which I was selected for, made me wonder where I had been for the last decade. Of course, there was the answer: ten years in a post, content, and therefore not knowing or caring what was happening on the other side of the fence.

The following are true reports of how some of my interviews went. I will name places only. One was on an estate near Wetherby where the farm manager conducted the verbal interview as we stood face to face in the farmyard. He then left me conversing with two other chaps, who were there for the same reason; then off he went into the farm office to give his report to his boss, a Captain somebody. When he returned, he told one of the chaps that he would not be suitable for the post, so that was the end of him. He then bade me to join him in his Land Rover, and off we flew around the estate at about 40 mph. His arms and hands were pointing in all directions: 'That's the wheat', 'That's the sugar beet', 'That's that' and 'This is this,' he announced, as though I had no idea what we were looking at.

When that was over and back at the yard, the bombshell came. He casually said, 'Oh, by the way, the wage is £10 less than the current rate.' I could not believe I was hearing this, and when I questioned the reason for it, let alone the legality of it, he came over so all double-faced and went on to say none of the others had complained, and how he thought this was how it was going to be in the future, and so on.

I didn't like his attitude, or the terms, therefore before he had the chance to tell me I too would not be suitable for the post, I informed him I no longer wished to be considered. I

*The Changing Face of Agriculture*

drove back home very disappointed, and in a way sad to learn that despite all that Agricultural Wages Board and the National Union of Allied and Agricultural Workers had done to achieve us a decent wage, there were men out there so desperate to get re-employed, they would go along with the former terms.

The next interview we attended in Leicestershire would have been more hilarious than disappointing, had it not been such a waste of time. We found the house we were to report to at 1 p.m. on a Saturday. It was quite a large house with two Mercs parked in its drive, which were badly in need of a good wash. As I approached the door I noticed it felt very quiet and bare, the hallmark of a bachelor's place. I went on to try three doors with no response, so after a while we asked at the nearest house if they knew the man. It turned out that the lady there was his housekeeper, and she assured us that he was in the house, but offered us no assistance, as she claimed to be off duty.

Again we tried to no avail. So we took a ride down the village and returned to try once more, now one hour behind the interview time, and sure enough this portly, tweed-suited, bedraggled-looking figure appeared at the door. He apologised humbly for having fallen to sleep after having his lunch, and added he had been so busy on his farm of late, although he gave me the impression that it was a long time ago. The post was advertised as being for an experienced tractor driver. He showed me two large dairy herds and little else. It appeared the post was mainly carting and pumping slurry and operating a diet feeder, with a little arable work thrown in. He picked his nose as he drove his car and didn't seem to care less whether he filled the post or not. We parted with, 'We will let you know,' and I am still waiting; this was 1980.

Later, I was called for an interview in West Yorkshire. This time I got to sit in a plush farm office in front of the farmer and his manager. The manager sat in silence, apart from when he was asked to comment. The boss looked through my application letter, and then his first question was, 'Are you a member of the NUAAW?'

'Yes, sir, I am,' I replied.

'Why?' he asked, sharply.

All I could think of to say was, 'Because I cannot afford not to be.'

He paused and said, 'Can you be more specific?'

Now, this was after the farm workers had been awarded the largest wage rise ever — I believe it was between £7 and £8 per week. But when the farmers, the Agricultural Wages Board and the NUAAW officials met, the farmers' side offered £3, and the hearing was dismissed. At a later date the case was re-heard and the original figure was agreed upon. So I thought, Here goes, and quoted this in my answer to his question and added, 'Without the union, three pounds it would have been.'

I shall never forget the look on the manager's face as I said those words. It revealed his inner turmoil, clearly condemning my bold utterance. However, he did say to the farmer, who was staring at him menacingly, 'I am afraid he is right, sir.'

The rest of the questions were only remotely relevant to the issue, in my view: Did I smoke? Did I drink? How many children had I? Were they good? What were my hobbies? And so on; need I say more…

I was never a staunch or militant union man, nor did I agree with all they said and did, but I did believe at one time or another they could be advantageous to a member, just as the National Farmers' Union could be to a farmer. Of course, I had some much more civil and realistic interviews where I met some very nice, understanding people, but the fact had to be faced that I was approaching fifty years of age, and the field of applicants was very full. The feeling of what one was getting was mutual, while the odds were largely in favour of the employer. I would go as far as to quote the maxim, 'It's not what you know, but who you know,' prevailed widely. But at the end of the day life goes on. There are many explanations as to why things had become as they were: mechanisation, technology, the state of the world's economy, farmers' costs and returns... some people even blamed the NUAAW for the slump in vacancies.

However, it did signify one thing, and that is that the issue had almost turned a full circle. For many years prior to this, it was the 'drift' from the land which was causing great alarm, as many statistics show that the drift from the land was a fact, and that the towns have been attracting labour from the farms for over one hundred years, wages being the dominant factor. Between June 1949 and June 1954, the number of male agricultural workers declined by 63,000.

Lieutenant Colonel M Lipton claimed in the *Farmer and Stock Breeder* on 13 July 1955 that we were heading for disaster unless the drift was checked, and pleaded for the abolition of National Service for land workers.

In 1955, a Conservative MP for Louth, Mr Cyril Osbourne, expressed his disbelief at an annual dinner. He referred to the weekly wage increase of eight shillings per week, making the minimum wage up to £6 15s 0d, as pitiable.

In 1957 the NUAAW lodged a claim for an increase of 19 shillings, bringing the weekly rate to £8, but the NFU opposed it, claiming it would cost a staggering £28.5 million.

Prior to, and for many years after the Agricultural Wages Regulations Act of 1924, the issue of workers' rights, wages and conditions have alarmed the farmers, the Ministry of Labour, MPs and other involved bodies. But even so, bearing in mind that I refer to the largest single industry in the country, positive results from any negotiations regarding the former in favour of the workers were slow to be forthcoming. In short, the agricultural worker is not, and never was, gullible to the point of cutting his nose off to spite his face; hence their tendency to watch the prosperity of the industry closely, and never rock the boat in a storm. So much for their loyalty.

Many events signify this, for in 1914 the weekly wage stood

at the grand total of 18 shillings. It rose during the 1914–1918 war and after it to a peak of 46s 10½d, only to slump back to 28 shillings when the Corn Regulations Act was repealed. So once again the farmer rose from his table, shook the cloth, and the workers received the crumbs – on, I must add, take it or leave it terms. Consequently, up to the 1939–1945 war the fear of unemployment governed any negotiations the workers' representatives might have envisaged on their behalf.

However, their loyalty paid off, for during the war a 100 per cent rise was achieved at once, taking the weekly wage from 35 shillings to 70 shillings, mainly down to a notable increase in output per farm worker and partly by substantial subsidies from the taxpayer. At last there was a major breakthrough for the British agricultural worker. The workers' side of the Wages Board had been discreetly scrutinising facts and figures.

When in 1960 they put their new claim for a reduction in the 47-hour week, plus a rise in the weekly wage to £8 10s, this was seen as ludicrous by the opposition. At the hearing, the workers' representatives gallantly quoted the Ministry of Labour's figures. They pointed out that agriculture had an annual turnover of £1,500 million, adding that the real output of the industry had increased by 29 per cent since 1947 on a steadily declining labour force, and claiming that the remaining workers had achieved an increase in output per worker of 70 per cent. Finally, they added that the £90 million increase in the wages bill was only 40 per cent above the 1949 level, making it a much less spectacular increase than the farmers' income.

This was a very well-put case, largely instigated by Mr Collinson, the NUAAW representative. Indeed, there was almost enough ammunition tendered there to slay Goliath. However, I am afraid it had to be conceded that the former results had a downside. While they were slowly winning the battle, which was their ultimate aim, to bring agricultural

## The Changing Face of Agriculture

workers' wages more in line with those of the town industries, the opposition retaliated by claiming that bigger wages and better terms for the workers would cost them their jobs. It appeared the drift was soon to be at an ebb, and even if it was not, it was no longer becoming a threat, because the line seemed to be that fewer workers were getting through just as much work. I would say this was largely due to the momentum of mechanisation, a trend that still seems to reduce even today's labour force.

When I look at some of the methods and ethics of arable farming these days, I feel glad to be out of it, although if farming is in one's blood it never goes away. Are these large intensive farming companies necessary? I think not, especially as they are largely responsible for the muddle-headed, bungled market strategies of today. It requires no skill or know-how to produce mountains of inferior products by artificial means, only to saturate the markets, which then require propping up by subsidies funded by the taxpayer. The whole system is so absurd. But where is the alternative? Perhaps if the man in the street discovered that all he needed was a brush and shovel to acquire money, he too would jump on the bandwagon.

Perhaps there is a flicker of hope now that this so-called organic farming is reappearing, but it must descend from the realms of cloud cuckoo land; no doubt Mr Santer will intervene in yet another field he knows nothing of, but at least if he can pronounce the word, I dare say it will be near enough.

Most of us are well enough aware that organic farming is nothing new. On the other hand many people know nothing about the ethics or the fundamental functions of it. It is grossly unfair and unjust to reintroduce organic farming in a haphazard and half-hearted manner. The recently published survey into egg production, for instance, proved to be way off the mark in many aspects. It made no secret of the fact that almost anything would pass as free-range eggs! A spokeswoman for one party said that the hens were more happy and more hygienically housed in battery cages, than cramped and confined in old barns and sheds etc. What a lot of nonsense! Until the hens are roaming in 30-acre fields (and the Lord knows there are plenty of those available these days), they are not producing free-range eggs.

The animals intended for slaughter are grown on longer to enhance maturity, and fed on a diet free from hormones, drugs, offal, concoctions and the like. Temic — a tried and tested insecticide used for the eradication of nematodes — is prohibited, or better still removed from the market. Similar strict legislation applied to all cropping. One cannot or should not regard the products from them as organic, otherwise the whole issue will be a complete farce before it gets off the ground, and furthermore yet another rip off for the consumer, i.e. the general public.

Frankly I have no confidence in the former issue. The cancer is too far embedded in the food chain to be resolved, therefore perhaps a ban on this lunacy would be more fitting. In any event, I feel it should be all or nothing. I suspect the chances of the return of the smaller producers are very remote, but in an ideal world they have a place.

The documentary pertaining to meat and its declining popularity, shown on BBC Channel 2 television on 2 May 1995, produced gallant attempts to paper over the cracks within the meat consumption issue. Whilst the narrator's words spoke volumes when he said that a few decades ago people needed no persuasion to eat meat. How true! However, they passed on the point at issue — why a vice versa situation has arisen.

The modern-minded boffin advocated that fancy packaging and new, unheard-of cuts, plus a little more mauling, would put the matter right. I suspect the decline and loss of confidence in meat on the whole goes much, much deeper. In a word, 'quality' has declined, and I make no apology for going on about it, for once one has had the real McCoy, so to speak, this mass-produced garbage we now have little option as to whether we purchase or not stands out like a sore thumb, and indeed signifies how real and threatening the relevance of BSE and salmonella is. My beliefs and theories on the issue have been even more strong, and indeed justified, since I, like the lady who took part in the second part of the programme, was a victim of salmonella food poisoning — and believe me I would not wish the ordeal on my worst enemy! It was perhaps some consolation when with the assistance of a doctor, an environmental health officer and an accurate account by myself of the contents of my meal prior to becoming ill, and naming the place of purchase of the product, which we all regarded as suspect, and the persistence of my solicitor, compensation was awarded. But even so, it was very annoying, frustrating, even embarrassing; for needless to say, my wife was involved, if not affected. Being a kitchen worker, she was banned from her job until the powers that be were satisfied with the analysis of the motion samples which we had to collect and tender every few days.

In all, it took five weeks to clear the matter up and certify us clear of the virus, or whatever. Incidentally there were a lot more people affected from the same source. The health officer visited the business in question and caught them red-handed with the remaining contaminated product refrigerated, which no doubt would have gone out for sale again.

We cannot pretend that incidents of this nature, of which there must be no shortage, do anyone any good. As to how the culprit was dealt with over the matter, I did not know; but the unsavoury publicity seemed to do the business no harm. Personally, I do not entirely blame the retailer for this malpractice. After all, they cannot sell gold when only tin is available to them. Therefore they are the pig in the middle, so to speak, for much of this virulence has lain unmalignant for many, many years, until the meddling of the scientists and agricultural boffins activated them — with dire results. Mass production is the root of all evil. It commandeers competition. The larger the business or concern, especially when privatised, the more corrupt, secretive, inept, inequitable, ineradicable, inessential, inconsiderate, inglorious and so on it becomes.

Take the bosses' pay, for example, in the privatised state utilities, claiming they work hard and are worthy of a £475,000 salary. Well, let's put it this way: would they, were they real bosses, pay an employee that kind of money? No way! It is not as though they possess any special skills that might justify this absurdity. As for the commodities they peddle — gas, water, electricity and telecoms, most people have at least three out of the four which are absolutely essential for one's everyday necessities. Not satisfied with this, they go on to dictate the terms on a take it or leave it basis.

There is no man or woman breathing worth that kind of money as a salary, no matter who they are, or what they do, other than a person conducting their own private business; then of course they are worthy of £475,000, if they are clever and astute enough to make it. The scales of balance are well ready for adjusting when these penny boardroom bosses' pay has increased by 82 per cent over the last five years, and the workers' pay a mere 15 per cent.

As regards a happy working relationship between employer and employee, I found the middle-sized farm less hectic and more rewarding. In my early days, the norm was to start out on a small farm and work up the ladder by gradualism. How different it has all become! Loyalty, pride in work status, terms of employment, honour, have long since gone out of the window, largely because the former play no integral part towards the making of money. My first post was a 50-acre farm where the boss and myself did all the work with no mechanical aid at all. It posed no stress, hassle, pressure, problems, hardships or the like whatsoever.

My last post was with a 6,000-acre chain of farms, where

the above evils prevailed daily, and where time and time again the maxim that the large farmer wasted more than the small one grew was proved to me. I can recall one incident of many which might shed some light on what I say here.

I was put in charge of irrigating a 100-acre block of King Edward potatoes, amongst other jobs. It was in the days when the irrigation pipes were several feet in length of alloy pipe, which one manually carried around the field when setting out the system and moving over the laterals in the crop. The system comprised 6" and 4" mains and laterals at intervals off them. It was jolly hard work trudging through the wet potato haulms, almost waist-deep when moving the laterals over, often in hot sunshine. However, this field of potatoes to which I refer had had several four-hour sessions of water through the summer, when we were ordered by the farm manager to pick up all the tackle and bring it out of the field. Again, no easy task, with each and every pipe to lift and stack onto specially adapted trailers. But alas, I could hardly believe what I was hearing, when on the Monday morning, the manager ordered us to set up the tackle in the same field and proceed. Now it would be a very naive man indeed not to be tempted to have a look at what was under those potato haulms, having been trudging through them for two months or so, and I knew that there was at least a 20-tons-per-acre crop already there, and the soil was becoming puddly wet.

Therefore I could not resist saying to the manager, with respect of course, that those potatoes did not require any more water. In fact, had they been my potatoes I would have concluded that to do so would have been signing the death warrant on them. He looked at me, and from his right profile, in a couldn't-care-less attitude, he said, 'Now, it's like this! The boss says those potatoes need more water. He tells me what to do and I tell you what to do, is that clear?'

So that was me put down. But I could not get the folly of their strategy out of my mind. For one thing, neither of them had lifted their fat posteriors from their car seats and actually walked into the field to have a look for themselves and see what the situation was. Let's not forget, it was by now well into September. Surely a man with a tiny IQ would have known that all it needed was the weather to break to plunge us into deep trouble. Sure enough, it happened. The rains started, causing a delay in the lifting of those potatoes. It went from bad to worse: the large machinery could not travel on the extra wet field; we reverted to hand picking, again under very adverse conditions, and it was a daunting task. There was a lot of crop damage through having to use 4x4 drive tractors churning about in the rows, and so on. Finally the crop was abandoned, needless to say resulting in a dead loss.

Now, in any right-thinking man's mind, the former drama was the result of extremely badly coordinated management and blatant pig-headedness. One of my old bosses used to say if a farmer needed a manager he had too much land, and I have to say I preferred the days when farm managers were not so numerous. In the days to which I refer, there was no such thing as a sinecure position, and I must concede that old habits die hard in my book.

Farm managers did not create their own popularity, they merely became more numerous due to fundamental national changes in agricultural policy. I suspect there are not many farm managers actually 'managing the money', as my old school of colleagues would put it. Therefore their role is not far removed from that of a good college-trained stockman or tractor driver; just as in bygone days the status of the head horseman and the foreman were almost equal. I have nothing personal against farm managers, but I do think it is a great pity that more of them do not ever have the opportunity or the resources to farm on their own.

Having worked closely with the boss for the first decade of my working life, I found it a totally different kettle of fish working under a manager. Some of them were arrogant, ruthless and self-centred, with an utter disregard for the other farm staff. Probably the cause of this behaviour would be due to the fact that after all they were only deputising for the boss role, and thus rendering themselves not sacred.

I saw and worked amongst some very good, practical farmers during the 1940s and 1950s, and I shall forever hold the memories and cherish the hospitality of those fine agriculturalists whom I had the pleasure of learning the fundamental rudiments of the game from, skills one cannot learn from a book, an agricultural college or any other source.

As I write, we are fast approaching a new millennium. I am not tired of living in the present one, despite all I have said; this world is still a beautiful place, full of beautiful things. It would not take much to put it right, albeit that's only my view. Perhaps the other 60 million people think it is right — who knows! But there is one thing I am sure of, and that is if one could choose between going back in time or forward in time, it would have to be back in time for me; for whatever the future holds for us, it cannot in my case supersede the life and times I have had.

Printed in the United Kingdom
by Lightning Source UK Ltd.
111081UKS00001B/57

9 781844 015849